Lecture Notes in Control and Information Sciences

Volume 479

This series aims to report new developments in the fields of control and information sciences—quickly, informally and at a high level. The type of material considered for publication includes:

1. Preliminary drafts of monographs and advanced textbooks
2. Lectures on a new field, or presenting a new angle on a classical field
3. Research reports
4. Reports of meetings, provided they are

   (a) of exceptional interest and
   (b) devoted to a specific topic. The timeliness of subject material is very important.

More information about this series at http://www.springer.com/series/642

Pauline Bernard

Observer Design for Nonlinear Systems

Springer

Pauline Bernard
Department of Electrical, Electronic,
and Information Engineering "Guglielmo
Marconi"
University of Bologna
Bologna, Italy

ISSN 0170-8643 ISSN 1610-7411 (electronic)
Lecture Notes in Control and Information Sciences
ISBN 978-3-030-11145-8 ISBN 978-3-030-11146-5 (eBook)
https://doi.org/10.1007/978-3-030-11146-5

Library of Congress Control Number: 2018966434

This Springer imprint is published by the registered company Springer Nature Switzerland AG
The registered company address is: Gewerbestrasse 11, 6330 Cham, Switzerland

À Laurent et Vincent, sans qui
ce livre n'aurait jamais vu le jour.

Preface

In many applications, estimating the current state of a dynamical system is crucial either to build a controller or simply to obtain real-time information on the system for decision-making or surveillance. A common way of addressing this problem is to place some sensors on/in the physical system and design an algorithm, called *observer*, whose role is to process the incomplete and imperfect information provided by the sensors and thereby construct a reliable estimate of the whole system state. Of course, such an algorithm can exist only if the measurements from the sensor somehow contain enough information to determine uniquely the state of the system; namely, the system is *observable*.

The number and quality of the sensors being often limited in practice due to cost and physical constraints, the observer plays a decisive role in a lot of applications. Many efforts have thus been made in the scientific community to develop universal methods for the construction of observers. Several conceptions of this object exist, but in this book, we mean by observer a finite-dimensional dynamical system fed with the measurements, and for which a function of the state must converge in time to the true system state. Although very satisfactory solutions are known for linear systems, nonlinear observer designs still suffer from a significant lack of generality. The very vast literature available on the subject consists of scattered results, each making specific assumptions on the structure and observability of the system. In other words, no unified and systematic method exists for the design of observers for nonlinear systems.

Actually, observer design may be more or less straightforward depending on the coordinates we choose to express the system dynamics. For instance, dynamics which seem nonlinear at first sight could turn out to be linear in other coordinates. Hence the importance of the choice of the coordinates for observer design. In particular, some specific structures, called *normal forms*, have been identified for allowing a direct and easier observer construction. One may cite for instance the state-affine forms with their so-called Luenberger or Kalman observers, or the triangular forms associated with the celebrated high-gain design. With this in mind, most solutions available in the literature actually fit in the following three-step methodology:

1. look for a reversible change of coordinates transforming the dynamics of the given nonlinear system into one of the identified normal forms,
2. design an observer in those new coordinates, and
3. deduce an estimate for the system state in the initial coordinates via inversion of the transformation.

Of course in order to follow this method, one needs to know

I. a list of normal forms and their associated observers,
II. under which conditions and thanks to which invertible transformation one can rewrite a dynamical system into one of those forms, and
III. how to compute the inverse of this transformation.

When browsing the literature, one discovers that the first two points have been extensively studied, although not always under this terminology. In fact, they constitute the core of the observer design problem and they are tightly linked since a particular form is of interest if it admits observers (Point I.) and if a large category of systems can be transformed into that form (Point II.). Therefore, Points I. and II. are often treated simultaneously. On the contrary, fewer results concern Point III., mainly because the observer problem is often considered theoretically solved, once an invertible transformation into a normal form has been found. But as we will see, this point is actually crucial in practice and is receiving increasing interest.

Based on those observations, the goal of this book is to gather the different designs existing in the literature, including the most recent ones, in a unified framework built around those three points. It is organized accordingly, namely in the following three parts:

Part I Normal forms and their observers,
Part II Transformation into a normal form, and
Part III Observer back into initial system coordinates

The book is intended to give the reader a good overview of the state-of-the-art in matters of observer design for nonlinear systems, with a focus on general design methods with global convergence. It gathers in a same framework, linearizations by output injection with Luenberger/Kalman designs, transformations into triangular forms and their high-gain designs, and transformations into a Hurwitz form for a nonlinear Luenberger design (also called Kazantzis–Kravaris design).

Writing this book would not have been possible without the guidance, expertise, everlasting enthusiasm (and so much more) of my Ph.D. advisors, Laurent Praly and Vincent Andrieu, to whom I want to express my deepest gratitude. A significant part of the results presented in this book is issued from our work together or from what they taught me. I would also like to warmly thank Alberto Isidori, who did me the honor of his presence in my Ph.D. committee and who encouraged me to transform my dissertation into a book.

Bologna, Italy
August 2018

Pauline Bernard

Contents

Chapter 1
Nonlinear Observability and the Observer Design Problem

This first chapter introduces the problem of observer design for nonlinear controlled systems and presents some basic notions of observability which will be needed throughout the book.

1.1 Observation Problem

We consider a general system of the form:

$$\dot{x} = f(x, u) \quad , \quad y = h(x, u) \tag{1.1}$$

with x the state in $\mathbb{R}^{d_x}$, u an input function with values in $\mathbb{R}^{d_u}$, y the output (or measurement) with values in $\mathbb{R}^{d_y}$, and f and h "sufficiently many times continuously differentiable"[1] functions defined on $\mathbb{R}^{d_x} \times \mathbb{R}^{d_u}$. We denote

- $X(x_0, t_0; t; u)$ the solution at time t of (1.1) with input u and passing through x_0 at time t_0. Most of the time, t_0 is the initial time 0 and x_0 the initial condition. In that case, we simply write $X(x_0; t; u)$.
- $Y(x_0, t_0; t; u)$ the output at time t of System (1.1) with input u passing through x_0 at time t_0, i.e.,

$$Y(x_0, t_0; t; u) = h(X(x_0, t_0; t; u), u(t)) .$$

To alleviate the notations when $t_0 = 0$, we simply note $y_{x_0,u}$, i.e.,

$$y_{x_0,u}(t) = h(X(x_0; t; u), u(t)) .$$

Those notations are used to highlight the dependency of the output on the initial condition (and the input). When this is unnecessary, we simply write $y(t)$.

[1]The need for differentiability and its order will vary locally throughout the book.

P. Bernard, *Observer Design for Nonlinear Systems*,
Lecture Notes in Control and Information Sciences 479,
https://doi.org/10.1007/978-3-030-11146-5_1

- $\mathscr{X}_0$ a subset of $\mathbb{R}^{d_x}$ containing the initial conditions that we consider for System (1.1). For any x_0 in $\mathscr{X}_0$, we denote $\sigma^+(x_0; u)$ (resp. $\sigma^+_{\mathscr{X}}(x_0; u)$) the maximal time of existence of $X(x_0; \cdot; u)$ in $\mathbb{R}^{d_x}$ (resp. in a set $\mathscr{X}$).
- $\mathscr{U}$ the set of all sufficiently many times differentiable inputs $u : [0, +\infty) \to \mathbb{R}^{d_u}$ which the system can be submitted to.
- U a subset of $\mathbb{R}^{d_u}$ containing all the values taken by the inputs $u \in \mathscr{U}$, i.e.,

$$\bigcup_{u \in \mathscr{U}} u([0, +\infty)) \subset U .$$

More generally, for an integer m such that any u in $\mathscr{U}$ is m times differentiable, $\overline{U}_m$ denotes a subset of $\mathbb{R}^{d_u(m+1)}$ containing the values taken by the inputs u in $\mathscr{U}$ and its first m derivatives, i.e.,

$$\bigcup_{u \in \mathscr{U}} \overline{u}_m([0, +\infty)) \subset \overline{U}_m ,$$

with $\overline{u}_m = (u, \dot{u}, \ldots, u^{(m)})$.

The object of this book is to address the following problem:

Observation problem *For any input u in $\mathscr{U}$, any initial condition x_0 in $\mathscr{X}_0$, find an estimate $\hat{x}(t)$ of $X(x_0; t; u)$ based on the only knowledge of the input and output up to time t, namely $u_{[0,t]}$ and $y_{[0,t]}$, and so that $\hat{x}(t)$ asymptotically approaches $X(x_0; t; u)$, at least when $\hat{x}(t)$ is defined on $[0, +\infty)$.*

Note that the solutions are defined from any points in $\mathbb{R}^{d_x}$, but we may choose to restrict our attention to those starting from a subset $\mathscr{X}_0$ of $\mathbb{R}^{d_x}$ (perhaps for physical reasons), and thus, we are only interested in estimating those particular solutions. Otherwise, take $\mathscr{X}_0 = \mathbb{R}^{d_x}$. As for the causality constraint that only the past values of the input $u_{[0,t]}$ can be used at time t, this may be relaxed in the case where the whole trajectory of u is known in advance, namely for a time-varying system.

The continuous differentiability of f says that any solution to System (1.1) is uniquely determined by its initial condition. Thus, the problem could be rephrased as: "given the input, find the only possible initial condition which could have produced the given output up to time t". Of course, this raises the question of uniqueness of the initial condition leading to a given output trajectory, at least after a certain time. This is related to the notion of observability which will be addressed later in this chapter. In any case, one could imagine simulating System (1.1) simultaneously for a set of initial conditions x_0 and progressively removing from the set those producing an output trajectory $Y(x_0; t; u)$ "too far" from $y(t)$ (with the notion of "far" to be defined). However, this method presents several drawbacks: First, one need to have a fairly precise idea of the initial condition to allow a trade-off between number of computations and estimation precision, and second, it heavily relies on the model (1.1) which could be imperfect. This path has nevertheless aroused a lot of research:

• either through stochastic approaches, adding random processes to the dynamics (1.1) and to the measurement, and following the probability distribution of the possible values of the state [6]

• or in a deterministic way, adding unknown admissible bounded disturbances to the dynamics (1.1) and to the measurement, and producing a "set-valued observer" or "interval observer" such as in [5, 8].

Another natural approach is the resolution of the minimization problem [11]

$$\hat{x}(t) = \mathsf{Argmin}_{\hat{x}} \int_0^t \left| Y(\hat{x}, t; \tau; u) - y(\tau) \right|^2 d\tau$$

or rather with finite memory

$$\hat{x}(t) = \mathsf{Argmin}_{\hat{x}} \int_{t-\bar{t}}^t \left| Y(\hat{x}, t; \tau; u) - y(\tau) \right|^2 d\tau \ .$$

Along this path, a first idea would be to integrate backward the differential equation (1.1) for a lot of initial conditions $\hat{x}$ at time t until $t - \bar{t}$ and select the "best" one, but this would require a huge number of computations which would be impossible to carry out online and, as before, it would rely too much on the model. Some methods have nonetheless been developed to alleviate the number of computations and solve this optimization problem online, in spite of its non-convexity and the presence of local minima (see [1] for a survey of existing algorithms).

In this book, the path we choose to follow is rather to look for a dynamical system using the current value of the output and the current (and past) values of the input, and whose state is guaranteed to provide (at least asymptotically) enough information to reconstruct the state of System (1.1). This dynamical system is called an *observer*. A more rigorous mathematical definition is the following (a sketch is given in Fig. 1.1).

Definition 1.1 An *observer* for System (1.1) initialized in $\mathcal{X}_0$ is a couple $(\mathcal{F}, \mathcal{T})$ where

- $\mathcal{F} : \mathbb{R}^{d_z} \times \mathbb{R}^{d_u} \times \mathbb{R}^{d_y} \to \mathbb{R}^{d_\xi}$ is continuous.
- $\mathcal{T}$ is a family of continuous functions $\mathcal{T}_u : \mathbb{R}^{d_z} \times [0, +\infty) \to \mathbb{R}^{d_x}$, indexed by u in $\mathcal{U}$, which respect the causality[2] condition:

$$\forall \tilde{u} : [0, +\infty) \to \mathbb{R}^{d_u}, \ \forall t \in [0, +\infty) \ , \ u_{[0,t]} = \tilde{u}_{[0,t]} \implies \mathcal{T}_u(\cdot, t) = \mathcal{T}_{\tilde{u}}(\cdot, t) \ .$$

- For any u in $\mathcal{U}$, any z_0 in $\mathbb{R}^{d_z}$, and any x_0 in $\mathcal{X}_0$ such that $\sigma^+(x_0; u) = +\infty$, any[3] solution $Z(z_0; t; u, y_{x_0,u})$ to

$$\dot{z} = \mathcal{F}(z, u, y_{x_0,u}) \tag{1.2}$$

[2]Again, this causality condition may be removed if the whole trajectory of u is explicitly known, for instance, in the case of a time-varying system where $u(t) = t$ for all t.

[3]We say "any solution" because $\mathcal{F}$ being only continuous, there may be several solutions. This is not a problem as long as any such solution verifies the required convergence property.

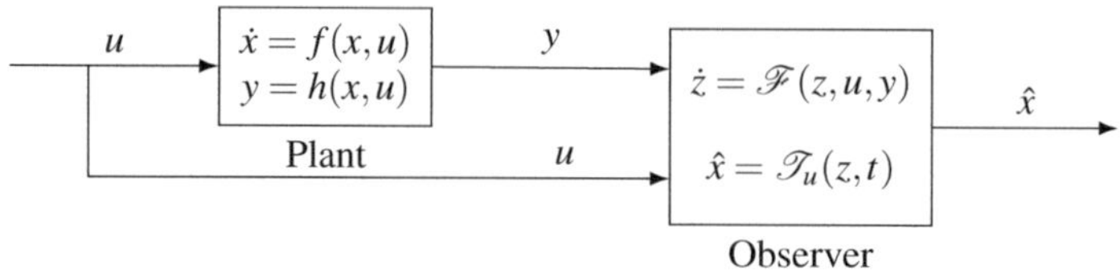

Fig. 1.1 Observer: dynamical system estimating the state of a plant from the knowledge of its output and input only

initialized at z_0 at time 0, with input u and $y_{x_0,u}$, exists on $[0, +\infty)$ and is such that

$$\lim_{t\to+\infty} \left|\hat{X}((x_0, z_0); t; u) - X(x_0; t; u)\right| = 0 \tag{1.3}$$

with

$$\hat{X}((x_0, z_0); t; u) = \mathscr{T}_u\left(Z(z_0; t; u, y_{x_0,u}), t\right) .$$

In other words, $\hat{X}((x_0, z_0); t; u)$ is an estimate of the current state of System (1.1) and the error made with this estimation asymptotically converges to 0 as time goes to the infinity.

If $\mathscr{T}_u$ is the same for any u in $\mathscr{U}$ and is defined on $\mathbb{R}^{d_\xi}$ instead of $\mathbb{R}^{d_\xi} \times \mathbb{R}$, i.e., is time-independent, $\mathscr{T}$ is said *stationary*. In this case, $\mathscr{T}$ directly refers to this unique function and we may simply say that

$$\dot{z} = \mathscr{F}(z, u, y) \quad , \quad \hat{x} = \mathscr{T}(z)$$

is an observer for System (1.1) initialized in $\mathscr{X}_0$.

In particular, we say that the observer is *in the given coordinates* if $\mathscr{T}$ is stationary and is a projection function from $\mathbb{R}^{d_z}$ to $\mathbb{R}^{d_x}$; namely, $\hat{X}((x_0, z_0); t; u)$ can be read directly from d_x components of $Z(z_0; t; u, y_{x_0,u})$. In the particular case where $d_x = d_z$ and $\mathscr{T}$ is the identity function, we may omit to precise $\mathscr{T}$.

Finally, when $\mathscr{X}_0 = \mathbb{R}^{d_x}$, i.e., the convergence is achieved for any initial condition of the system, we say "observer" without specifying $\mathscr{X}_0$.

Remark 1.1 We will see in Chap. 4 that it is sometimes useful to write the observer dynamics (1.2) as a differential inclusion. In this case, $\mathscr{F}$ is a set-valued map and everything else remains unchanged.

The time dependence of $\mathscr{T}_u$ enables to cover the case where the knowledge of the input and/or the output is used to build the estimate $\hat{x}$ from the observer state z. In particular, using the output sometimes enables to reduce the dimension of the observer state (and thus alleviate the computations), thus obtaining a *reduced-order observer*. For instance, in the so-called *immersion and invariance* approach ("I&I") developed in [3, 7], the state x is separated in a measured part x_m and a non-measured part $x_{nm} \in \mathbb{R}^{d_x - d_y}$ and the dynamics of z are chosen to make a time-varying set

$$\{(z, x_m, x_{nm}) \in \mathbb{R}^{d_z} \times \mathbb{R}^{d_y} \times \mathbb{R}^{d_x - d_y} \ : \ \beta(z, x_m, t) = \phi(x_{nm})\}$$

invariant and asymptotically stable (uniformly in time) with appropriately chosen functions β and ϕ. If ϕ is left invertible, replacing x_m by the measurement y, this enables to recover the non-measured part x_{nm} of the state, i.e.,

$$\mathscr{T}(z, t) = (y(t), \phi^*(\beta(z, y(t), t)))$$

with ϕ^* a globally defined left inverse of ϕ. The advantage is that since x_m is not estimated, the observer state is of smaller dimension, but with the counterpart that the estimate $\hat{x}$ depends directly on y and is therefore affected by measurement noise. This kind of observer will not be studied in this book, and the map $\mathscr{T}$ will not depend on the output y. On the other hand, we will see that it is sometimes necessary to use the input (either implicitly or explicitly) in $\mathscr{T}_u$, but always keeping in mind the causality condition.

The advantage of having an observer in the given coordinates is that the estimate of the system state can directly be read from the observer state. This spares the maybe-complicated computation of $\mathscr{T}_u$. Writing the dynamics of the observer in the given coordinates constitutes one of the goals of this book, but we will see that, unfortunately, it is not always possible, nor easy.

Anyhow, the role of an observer is to estimate the system state based on the knowledge of the input and output. This means that those signals somehow contain enough information to determine uniquely the whole state of the system. This brings us to the notion of observability.

1.2 Observability and Observer Design for Nonlinear Systems

1.2.1 Some Notions of Observability

In order to have an observer, a detectability property must be satisfied:

Lemma 1.1 *Assume there exists an observer for System* (1.1). *Then, System* (1.1) *is detectable for any u in $\mathscr{U}$; i.e., for any u in $\mathscr{U}$ and for any (x_a, x_b) in $\mathscr{X}_0 \times \mathscr{X}_0$ such that $\sigma^+(x_a, u) = \sigma^+(x_b, u) = +\infty$ and*

$$y_{x_a,u}(t) = y_{x_b,u}(t) \quad \forall t \geq 0 \,,$$

we have

$$\lim_{t \to \infty} |X(x_a; t; u) - X(x_b; t; u)| = 0 \,.$$

The property of detectability says that even if two different initial conditions are not distinguishable with the output, the corresponding system solutions become close asymptotically, and thus, we still get a "good" estimate no matter which we pick. This is a well-known necessary condition which can be found, for instance, in [2] and which admits the following straightforward proof.

Proof Consider any u in $\mathscr{U}$ and any (x_a, x_b) in $\mathscr{X}_0^2$ such that $\sigma^+(x_a, u) = \sigma^+(x_b, u) = +\infty$ and $y_{x_a,u} = y_{x_b,u}$. Take z_0 in $\mathbb{R}^{d_z}$, and pick a solution $Z(z_0; t; u; y_{x_a,u})$ of (1.2) with input $y_{x_a,u}$. It is also a solution to (1.2) with input $y_{x_b,u}$. Therefore, by denoting $\hat{X}((x_a, z_0); t; u) = \mathscr{T}\big(Z(z_0; t; u, y_{x_a,u}), u(t), y_{x_a,u}(t)\big)$, we have

$$\lim_{t\to\infty} |\hat{X}((x_a, z_0); t; u) - X(x_a; t; u)| = 0$$

and

$$\lim_{t\to\infty} |\hat{X}((x_a, z_0); t; u) - X(x_b; t; u)| = 0\ .$$

The conclusion follows. □

This means that detectability at least is necessary to be able to construct an observer. Actually, we often ask for stronger observability properties such as:

Definition 1.2 Consider an open subset $\mathscr{S}$ of $\mathbb{R}^{d_x}$. System (1.1) is

- *Distinguishable* on $\mathscr{S}$ for some input $u : \mathbb{R} \to \mathbb{R}^{d_u}$ if for all (x_a, x_b) in $\mathscr{S} \times \mathscr{S}$,

$$y_{x_a,u}(t) = y_{x_b,u}(t) \quad \forall t \in [0, \min\{\sigma^+(x_a; u), \sigma^+(x_b; u)\}) \quad \Longrightarrow \quad x_a = x_b\ .$$

- *Instantaneously distinguishable* on $\mathscr{S}$ for some input $u : \mathbb{R} \to \mathbb{R}^{d_u}$ if for all (x_a, x_b) in $\mathscr{S} \times \mathscr{S}$, for all $\bar{t}$ in $(0, \min\{\sigma^+(x_a; u), \sigma^+(x_b; u)\})$

$$y_{x_a,u}(t) = y_{x_b,u}(t) \quad \forall t \in [0, \bar{t}) \quad \Longrightarrow \quad x_a = x_b\ .$$

- *Uniformly observable* on $\mathscr{S}$ if it is distinguishable on $\mathscr{S}$ for any input $u : \mathbb{R} \to \mathbb{R}^{d_u}$ (not only for u in $\mathscr{U}$).
- *Uniformly instantaneously observable* on $\mathscr{S}$ if it is instantaneously distinguishable on $\mathscr{S}$ for any input $u : \mathbb{R} \to \mathbb{R}^{d_u}$ (not only for u in $\mathscr{U}$).

In particular, the notion of instantaneous distinguishability means that the state of the system can be uniquely deduced from the output of the system as quickly as we want. In the particular case, where f, h, and u are analytical, y is an analytical function of time [4, Sect. 10.5.3] and the notions of distinguishability and instantaneous distinguishability are equivalent because two analytical functions which are equal on an interval are necessarily equal on their maximal interval of definition. Besides, for any x_0, there exists t_{x_0} such that

$$y_{x_0,u}(t) = \sum_{k=0}^{+\infty} \frac{y_{x_0,u}^{(k)}(0)}{k!} t^k \quad , \quad \forall t \in [0, t_{x_0}) \ ,$$

and distinguishability is thus closely related to the important notion of differential observability which will be defined in Chap. 5 and which roughly says that the state of the system at a specific time is uniquely determined by the value of the output and of its derivatives (up to a certain order) at that time.

The notion of uniform observability could appear unnecessary at first sight because it seems sufficient that the system is observable for any u in $\mathscr{U}$, namely for any considered input, rather than for any $u : \mathbb{R} \to \mathbb{R}^{d_u}$. However, we will see that this (strong) observability property infers some structural properties on the system which are useful for the design of certain observers.

In fact, more or less strong observability properties are needed depending on the observer design method and on what is required from the observer (tunability, exponential convergence, etc.). For example, it is shown in [2] that for autonomous systems, instantaneous distinguishability is necessary to have a tunable observer, i.e., an observer giving an arbitrarily small error on the estimate in an arbitrarily short time.

1.2.2 Observer Design

It is proved in [2] that if there exists an observer $(\mathscr{F}, \mathscr{T})$ for an autonomous system

$$\dot{x} = f(x) \quad , \quad y = h(x)$$

and a compact subset of $\mathbb{R}^{d_x} \times \mathbb{R}^{d_z}$ which is invariant by the dynamics $(f, \mathscr{F})$, then there exist compact subsets $\mathscr{C}_x$ of $\mathbb{R}^{d_x}$ and $\mathscr{C}_z$ of $\mathbb{R}^{d_z}$, and a closed set-valued map T defined on $\mathscr{C}_x$ such that the set

$$\mathscr{E} = \{(x, z) \in \mathscr{C}_x \times \mathscr{C}_z \ : \ z \in T(x)\}$$

is invariant, attractive, and verifies:

$$\forall (x, z) \in \mathscr{E} \ , \quad \mathscr{T}(z, h(x)) = x \ .$$

In other words, the pair made of the system state x (following the dynamics f) and the observer state z (following the dynamics $\mathscr{F}$) converges necessarily to the graph of some set-valued map T and $\mathscr{T}$ is a left inverse of this mapping. Note that this injectivity is of a peculiar kind since it is conditional to the knowledge of the output; namely, "$x \mapsto T(x)$ is injective knowing $h(x)$". This result justifies the usual methodology of observer design for autonomous systems which consists in transforming, via a function T, the system into a form for which an observer is available, then design the observer in those new coordinates (i.e., find $\mathscr{F}$), and finally deduce an estimate in the original coordinates via inversion of T (i.e., find $\mathscr{T}$). Note that in practice, we look for a single-valued map T because it is simpler to manipulate than a set-valued map.

When considering a time-varying or controlled system, the same methodology can be used, but two paths are possible:

- Either we keep looking for a stationary transformation $x \mapsto T(x)$ like for autonomous systems.
- Or we look for a time-varying transformation $(x,t) \mapsto T_u(x,t)$ which depends either explicitly or implicitly on the input u.

It is actually interesting to detail what we mean by explicitly/implicitly. In building a time-varying transformation, two approaches exist, each attached to a different vision of controlled systems:

- Either we consider, as in System (1.1), that only the current value of the input (or sometimes the extended input $\overline{u}_m = (u, \dot{u}, \ldots, u^{(m)})$) is necessary to determine $T_u(\cdot,t)$ at time t; i.e., there exists a function $\tilde{T}$ such that for any u in $\mathscr{U}$, $T_u(x,t) = \tilde{T}(x, u(t))$.
- Or we consider System (1.1) as a family of systems indexed by u in $\mathscr{U}$, i.e.,

  $$\dot{x} = f_u(x,t) \quad , \quad y = h_u(x,t)$$

  and we obtain a family of functions T_u, each depending on trajectory of u in $\mathscr{U}$. In this case, it is necessary to ensure that $T_u(\cdot,t)$ depends only on the past values of u to guarantee causality.

Along this book, we will encounter/develop methods from each of those categories. In any case, here is a sufficient condition to build an observer for System (1.1):

Theorem 1.1 *Consider an integer d_ξ and continuous maps $F : \mathbb{R}^{d_\xi} \times \mathbb{R}^{d_u} \times \mathbb{R}^{d_y} \to \mathbb{R}^{d_\xi}$, $H : \mathbb{R}^{d_\xi} \times \mathbb{R}^{d_u} \to \mathbb{R}^{d_y}$ and $\mathscr{F} : \mathbb{R}^{d_\xi} \times \mathbb{R}^{d_u} \times \mathbb{R}^{d_y} \to \mathbb{R}^{d_\xi}$ such that*

$$\dot{\hat{\xi}} = \mathscr{F}(\hat{\xi}, u, \tilde{y}) \tag{1.4}$$

is an observer for[4]

$$\dot{\xi} = F(\xi, u, H(\xi, u)) \quad , \quad \tilde{y} = H(\xi, u), \tag{1.5}$$

that is, for any $(\hat{\xi}_0, \xi_0)$ in $(\mathbb{R}^{d_\xi})^2$ and any u in $\mathscr{U}$, any solution $\hat{\Xi}(\hat{\xi}_0; t; u, \tilde{y}_{\xi_0,u})$ of (1.4) *and any solution $\Xi(\xi_0; t; u)$ of* (1.5) *verify*

$$\lim_{t \to +\infty} \left| \hat{\Xi}(\hat{\xi}_0; t; u, \tilde{y}_{\xi_0,u}) - \Xi(\xi_0; t; u) \right| = 0\,. \tag{1.6}$$

---

[4]The expression of the dynamics under the form $F(\xi, u, H(\xi, u))$ can appear strange and abusive at this point because it is highly non-unique, and we should rather write $F(\xi, u)$. However, we will see in Part I how specific structures of dynamics $F(\xi, u, y)$ allow the design of an observer (1.4).

Now suppose that for any u in $\mathcal{U}$, there exists a continuous function $T_u : \mathbb{R}^{d_x} \times \mathbb{R} \to \mathbb{R}^{d_\xi}$ and a subset $\mathcal{X}$ of $\mathbb{R}^{d_x}$ such that:

(a) For any x_0 in $\mathcal{X}_0$ such that $\sigma^+(x_0; u) = +\infty$, $X(x_0; \cdot; u)$ remains in $\mathcal{X}$.
(b) There exists a concave $\mathcal{K}$ function ρ and a positive real number $\bar{t}$ such that for all (x_a, x_b) in $\mathcal{X}^2$ and all $t \geq \bar{t}$

$$|x_a - x_b| \leq \rho\Big(|T_u(x_a, t) - T_u(x_b, t)|\Big) ,$$

i.e., $x \mapsto T_u(x, t)$ becomes injective on $\mathcal{X}$, uniformly in time and in space, after a certain time $\bar{t}$.
(c) T_u transforms System (1.1) into System (1.5), i.e., for all x in $\mathcal{X}$ and all t in $[0, +\infty)$

$$L_{(f,1)}T_u(x, t) = F(T_u(x, t), u(t), h(x, u(t))) \quad , \quad h(x, u(t)) = H(T_u(x, t), u(t)) , \tag{1.7}$$

where $L_{(f,1)}T_u$ is the Lie derivative of T_u along the extended vector field $(f, 1)$, namely

$$L_{(f,1)}T_u(x, t) = \lim_{h\to 0} \frac{T_u(X(x, t; t + h; u), t + h) - T_u(x, t)}{h}$$

(d) T_u respects the causality condition

$$\forall \tilde{u} : [0, +\infty) \to \mathbb{R}^{d_u},\ \forall t \in [0, +\infty) ,\ u_{[0,t]} = \tilde{u}_{[0,t]} \implies T_u(\cdot, t) = T_{\tilde{u}}(\cdot, t) .$$

Then, for any u in $\mathcal{U}$, there exists a function $\mathcal{T}_u : \mathbb{R}^{d_\xi} \times [0, +\infty) \to \mathbb{R}^{d_x}$ (verifying the causality condition) such that for each $t \geq \bar{t}$, $\xi \mapsto \mathcal{T}_u(\xi, t)$ is uniformly continuous on $\mathbb{R}^{d_\xi}$ and verifies

$$\mathcal{T}_u(T_u(x, t), t) = x \quad \forall x \in \mathcal{X} .$$

Besides, denoting $\mathcal{T}$ the family of functions $\mathcal{T}_u$ for u in $\mathcal{U}$, $(\mathcal{F}, \mathcal{T})$ is an observer for System (1.1) initialized in $\mathcal{X}_0$.

Solving the partial differential equation (1.7) a priori gives a solution T_u depending on the whole trajectory of u and rather situates this result in the last design category presented above. But this formalism actually covers all three approaches and was chosen for its generality. In fact, the dependence of T_u on u may vary, but what is crucial is that they all transform the system into the same target form (1.5) for which an observer (1.4) is known.

Proof Take u in $\mathcal{U}$. For any $t \geq \bar{t}$, $x \mapsto T_u(x, t)$ is injective on $\mathcal{X}$; thus, there exists a function $T_{u,t}^{-1} : T_u(\mathcal{X}, t) \to \mathcal{X}$ such that for all x in $\mathcal{X}$, $T_{u,t}^{-1}(T_u(x, t)) = x$. Taking any $\tilde{u} : [0, +\infty) \to \mathbb{R}^{d_u}$ such that $u_{[0,t]} = \tilde{u}_{[0,t]}$ thus gives $T_{u,t}^{-1} = T_{\tilde{u},t}^{-1}$ on $T_u(\mathcal{X}, t) = T_{\tilde{u}}(\mathcal{X}, t)$ according to d). Besides, with b), for all (ξ_1, ξ_2) in $T_u(\mathcal{X}, t)^2$,

$$|T_{u,t}^{-1}(\xi_1) - T_{u,t}^{-1}(\xi_2)| \leq \rho\Big(|\xi_1 - \xi_2|\Big) . \tag{1.8}$$

Applying [9, Theorem 2] to each component of $T_{u,t}^{-1}$, there exist $c > 0$ and an extension[5] of $T_{u,t}^{-1}$ on $\mathbb{R}^{d_\xi}$ verifying (1.8) with $\bar{\rho} = c\rho$ for all (ξ_1, ξ_2) in $(\mathbb{R}^{d_\xi})^2$ (i.e., $T_{u,t}^{-1}$ is uniformly continuous on $\mathbb{R}^{d_\xi}$) and such that $T_{u,t}^{-1} = T_{\tilde{u},t}^{-1}$ on $\mathbb{R}^{d_\xi}$. Defining $\mathscr{T}$ on $\mathbb{R}^{d_\xi} \times [0, +\infty)$ as

$$\mathscr{T}_u(\xi, t) = \begin{cases} T_{u,t}^{-1}(\xi) & , \text{ if } t \geq \bar{t} \\ 0 & , \text{ otherwise} \end{cases}$$

$\mathscr{T}_u$ verifies the causality condition and we have for all $t \geq \bar{t}$ and all (x, ξ) in $\mathscr{X} \times \mathbb{R}^{d_\xi}$,

$$|\mathscr{T}_u(\xi, t) - x| \leq \bar{\rho}\Big(|\xi - T_u(x, t)|\Big) . \tag{1.9}$$

Now consider x_0 in $\mathscr{X}_0$ such that $\sigma^+(x_0; u) = +\infty$. Then, from a) and c), since $X(x_0; \cdot; u)$ remains in $\mathscr{X}$ and $T_u(X(x_0; \cdot; u), t)$ is a solution to (1.5) initialized at $\xi_0 = T_u(x_0, 0)$ and for all t, $y_{x_0,u}(t) = \tilde{y}_{\xi_0,u}(t)$. Thus, because of (1.6), for any $\hat{\xi}_0$ in $\mathbb{R}^{d_\xi}$ and any solution $\hat{\Xi}(\hat{\xi}_0; t; u, y_{x_0,u})$ of

$$\dot{\hat{\xi}} = \mathscr{F}(\hat{\xi}, u, y_{x_0,u})$$

we have

$$\lim_{t \to +\infty} \Big|\hat{\Xi}(\hat{\xi}_0; t; u, y_{x_0,u}) - T_u(X(x_0; t; u), t)\Big| = 0 .$$

If follows from (1.9) that

$$\lim_{t \to +\infty} \Big|\hat{X}((x_0, \hat{\xi}_0); t; u) - X(x_0; t; u)\Big| = 0$$

with $\hat{X}((x_0, \hat{\xi}_0); t; u) = \mathscr{T}_u(\hat{\Xi}(\hat{\xi}_0; t; u, y_{x_0,u}), t)$. Thus, $(\mathscr{F}, \mathscr{T})$ is an observer for System (1.1). □

Remark 1.2 Without the assumption of concavity of ρ, it is still possible to show that $x \mapsto T_u(x, t)$ admits a continuous left inverse $\mathscr{T}_u$ defined on $\mathbb{R}^{d_\xi}$. But, as shown in [10, Example 4], continuity of $\mathscr{T}$ is not enough to deduce the convergence of $\hat{x}$ from that of $\hat{\xi}$: Uniform continuity[6] is necessary. Note that if $\mathscr{X}$ is bounded, the concavity of ρ is no longer a constraint, since a concave upper approximation can always be obtained by saturation of ρ (see [9] for more details).

---

[5]Denoting $T_{u,t,j}^{-1}$ the jth component of $T_{u,t}^{-1}$, take $T_{u,t,j}^{-1}(\xi) = \min_{\tilde{\xi} \in T_u(\mathscr{X},t)} \{T_{u,t,j}^{-1}(\tilde{\xi}) + \rho(|\tilde{\xi} - \xi|)\}$ or equivalently $T_{u,t,j}^{-1}(\xi) = \min_{x \in \mathscr{X}} \{x_j + \rho(|T_u(x, t) - \xi|)\}$

[6]A function γ is uniformly continuous if and only if $\lim_{n \to +\infty} |x_n - y_n| = 0$ implies $\lim_{n \to +\infty} |\gamma(x_n) - \gamma(y_n)| = 0$. This property is indeed needed in the context of observer design.

Besides, if there exists a compact set $\mathcal{C}$ such that $\mathcal{X}$ is contained in $\mathcal{C}$, it is enough to ensure the existence of ρ for (x_a, x_b) in $\mathcal{C}^2$. As long as for all t, $x \mapsto T_u(x,t)$ is injective on $\mathcal{C}$, then for all t, there exists a concave $\mathcal{K}$ function ρ_t verifying the required inequality for all (x_a, x_b) in $\mathcal{C}^2$ (see Lemma A.9). Thus, only uniformity in time should be checked, namely that there exists a concave $\mathcal{K}$ function ρ greater than all the ρ_t; in other words, that $x \mapsto T(x,t)$ does not become "less and less injective" with time. Of course, when T_u is time-independent, no such problem exists and it is sufficient to have $x \mapsto T_u(x)$ injective on $\mathcal{C}$. This is made precise in the following corollary.

Corollary 1.1 *Consider an integer* d_ξ *and continuous maps* $F : \mathbb{R}^{d_\xi} \times \mathbb{R}^{d_u} \times \mathbb{R}^{d_y} \to \mathbb{R}^{d_\xi}$, $H : \mathbb{R}^{d_\xi} \times \mathbb{R}^{d_u} \to \mathbb{R}^{d_y}$ *and* $\mathcal{F} : \mathbb{R}^{d_\xi} \times \mathbb{R}^{d_u} \times \mathbb{R}^{d_y} \to \mathbb{R}^{d_\xi}$ *such that* (1.4) *is an observer for* (1.5). *Suppose there exists a continuous function* $T : \mathbb{R}^{d_x} \to \mathbb{R}^{d_\xi}$ *and a compact set* $\mathcal{C}$ *of* $\mathbb{R}^{d_x}$ *such that:*

- *For any* x_0 *in* $\mathcal{X}_0$ *such that* $\sigma^+(x_0, u) = +\infty$, $X(x_0; \cdot; u)$ *remains in* $\mathcal{C}$.
- $x \mapsto T(x)$ *is injective on* $\mathcal{C}$.
- T *transforms System* (1.1) *into System* (1.5) *on* $\mathcal{C}$, *i.e., for all* x *in* $\mathcal{C}$, *all* u *in* $\mathcal{U}$, *all* t *in* $[0, +\infty)$

$$L_{f(\cdot,u)}T(x) = F(T(x), u(t), h(x, u(t))) \quad , \quad h(x, u(t)) = H(T(x), u(t)) \, .$$

Then, there exists a uniformly continuous function $\mathcal{T} : \mathbb{R}^{d_\xi} \to \mathbb{R}^{d_x}$ *such that*

$$\mathcal{T}(T(x)) = x \quad \forall x \in \mathcal{C} \, ,$$

and $(\mathcal{F}, \mathcal{T})$ *is an observer for System* (1.1) *initialized in* $\mathcal{X}_0$.

Proof This is a direct consequence of Lemma A.9 and Theorem 1.1. □

1.3 Organization of the Book

As illustrated in Fig. 1.2, Theorem 1.1 shows that a possible strategy to design an observer is to transform the system into a favorable form (1.5) for which an observer is known, and then bring the estimate back into the initial coordinates by inverting the transformation. This design procedure is widely used in the literature and raises three crucial questions:

1. Which normal forms (1.5) do we know and which observers are they associated to?
2. How to transform a given nonlinear system into one of those forms?
3. How to invert the transformation?

The present book addresses each of those questions and is thus organized accordingly, dedicating one part to each of them. Here is a more detailed account of the book's contents:

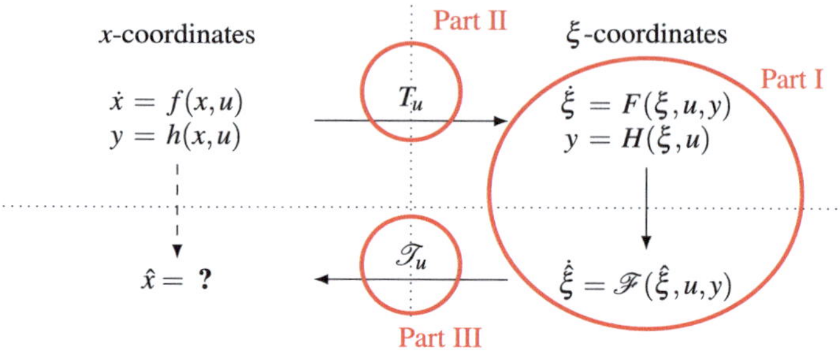

Fig. 1.2 Process of observer design suggested by Theorem 1.1 and organization of the book

Part I: Normal forms and their observers. We start by making a list of system structures (1.5) for which we know an observer (1.4). We call those favorable structures *normal forms*. Chapter 3 first reviews the state-affine normal forms with their associated Luenberger or Kalman observers. Then, Chap. 4 concentrates on triangular forms with high-gain, homogeneous, or mixed high-gain Kalman observers.

Part II: Transformation into a normal form. This part addresses the problem of transforming a nonlinear system into each of the previously mentioned normal forms. In each case, sufficient observability conditions on the system are given. In Chap. 6, we start by studying how a nonlinear system can be transformed into a state-affine form, with the so-called linearization problem or the more general nonlinear Luenberger design. Then, in Chap. 7, the question of transforming a system into a triangular form is thoroughly investigated.

Part III: Expression of the observer dynamics in the initial system coordinates. Although the observer design problem seems solved with Part I and II according to Theorem 1.1, implementation issues may arise such as the computation of the inverse $\mathscr{T}_u$ of the transformation. That is why, in Part III, we account for a recently developed methodology whose goal is to avoid the inversion of T_u by bringing the dynamics (1.4) back in the x-coordinates; i.e., find $\dot{\hat{x}}$ and obtain an observer *in the given coordinates* as defined in Definition 1.1. Although this process is quite common in the case where T_u is a diffeomorphism, completeness of solutions is not always ensured and we show how to solve this problem. Most importantly, this method extends to the more complex situation where T_u is only an injective immersion; i.e., the dimension of the observer state is larger than the one of the system states. This is done by adding some new coordinates to the system.

References

1. Alamir, M.: Nonlinear Observers and Applications. Lecture notes in control and information sciences, vol. 363, pp. 139–179. Springer, Berlin (2007). Chapter Nonlinear Moving Horizon Observers: Theory and Real-Time Implementation

2. Andrieu, V., Besançon, G., Serres, U.: Observability necessary conditions for the existence of observers. In: IEEE Conference on Decision and Control (2013)
3. Astolfi, A., Ortega, R.: Immersion and invariance : a new tool for stabilization and adaptive control of nonlinear systems. IEEE Trans. Autom. Control **48**(4) (2003)
4. Dieudonné, J.: Foundations of Modern Analysis. Academic, New York (1960)
5. Gouzé, J., Rapaport, A., Hady-Sadok, M.: Interval observers for uncertain biological systems. Ecol. Model. **133**, 45–56 (2000)
6. Jazwinski, A.H.: Stochastic Processes and Filtering Theory. Academic, New York (1970)
7. Karagiannis, D., Astolfi, A.: Nonlinear observer design using invariant manifolds. In: IEEE Conference on Decision and Control and European Control Conference (2005)
8. Lin, H., Zhai, G., Antsaklis, P.J.: Set-valued observer design for a class of uncertain linear systems with persistent disturbance. In: American Control Conference (2003)
9. McShane, E.J.: Extension of range of functions. Bull. Am. Math. Soc. **40**(12), 837–842 (1934)
10. Shim, H., Liberzon, D.: Nonlinear observers robust to measurement disturbances in an ISS sense. IEEE Trans Autom Control **61**(1) (2016)
11. Zimmer, G.: State observation by on-line minimization. Int J Control **60**(4), 595–606 (1994)

Part I
Normal Forms and Their Observers

Chapter 2
Introduction

In this part, we consider systems of the form[1]

$$\dot{\xi} = F(\xi, u, y) \quad , \quad y = H(\xi, u) \tag{2.1}$$

with ξ the state in $\mathbb{R}^{d_\xi}$, u an input with values in $U \subset \mathbb{R}^{d_u}$, y the output with values in $\mathbb{R}^{d_y}$ and F (resp. H) a continuous function defined on $\mathbb{R}^{d_\xi} \times \mathbb{R}^{d_u} \times \mathbb{R}^{d_y}$ (resp. $\mathbb{R}^{d_\xi} \times \mathbb{R}^{d_u}$). We are interested in finding normal forms, namely specific expressions of the functions F and H such that an explicit observer for System (2.1) can be written directly in the ξ-coordinates. Indeed, an a priori knowledge of such forms is necessary to apply Theorem 1.1 and design an observer for a nonlinear system.

We do not claim to be exhaustive, neither about the list of normal forms nor about their history. But we select the most popular and general forms and associated observers, and endeavor to give the most sensible references. Note that according to Theorem 1.1, we are only interested in global observers with guaranteed convergence. This excludes, for example, the extended Kalman filters obtained by linearizing the dynamics and the output along the trajectory of the estimate [3]. Indeed, their convergence is only local in the sense that the estimate converges to the true state if the initial error is not too large and the linearization does not present any singularity [1] and references therein.

The normal forms and their observers covered in this part are summed up in Table 2.1. We will need the following definition:

Definition 2.1 For a linear time-varying system of the form

$$\dot{\chi} = A(\nu)\chi \quad , \quad y = C(\nu)\chi \tag{2.2}$$

with input ν and output y, we define:

[1]The notation $F(\xi, u, y)$ is somehow abusive because y is not an input to the dynamics of ξ. We should rather write $F(\xi, u, H(\xi, u))$ as in (1.5), but this latter notation is less straight-forward. We thus decided to keep the former for clarity.

P. Bernard, *Observer Design for Nonlinear Systems*,
Lecture Notes in Control and Information Sciences 479,
https://doi.org/10.1007/978-3-030-11146-5_2

Table 2.1 Normal forms and their associated observer design

Structure			Observability assumption	Observer design
State-affine forms	H nonlinear	A constant Hurwitz	∅	Copy of the dynamics
	H linear	A and C constant	(A, C) observable	Luenberger
		A or C non constant A bounded	(u, y) regularly persistent	Kalman
Triangular forms	Nominal (4.2)	Φ_i bszLipschitz	∅	High-gain
		Φ_i Hölder (4.6), $d_0 \in [-1, 0]$	∅	Homogeneous of degree d_0
		Φ_i continuous and bounded trajectories/input	∅	Cascade of homogeneous observers of degree -1
	General (4.14)	Φ_i Lipschitz A_i bounded C bounded	(u, y) locally regular	High-gain Kalman

- the *transition matrix*[2] Ψ_ν as the unique solution to

$$\frac{\partial \Psi_\nu}{\partial \tau}(\tau, t) = A(\nu(\tau))\Psi_\nu(\tau, t)$$
$$\Psi_\nu(t, t) = I \ .$$

- the *observability grammian* as the function defined by:

$$\Gamma_\nu(t_0, t_1) = \int_{t_0}^{t_1} \Psi_\nu(\tau, t_0)^\top C(\nu(\tau))^\top C(\nu(\tau))\Psi_\nu(\tau, t_0)\, d\tau$$

- the *backward observability grammian* as the function defined by:

$$\Gamma_\nu^b(t_0, t_1) = \int_{t_0}^{t_1} \Psi_\nu(\tau, t_1)^\top C(\nu(\tau))^\top C(\nu(\tau))\Psi_\nu(\tau, t_1)\, d\tau$$

The transition matrix is used to express the solutions to system (2.2) because it verifies

$$\chi(\chi_0, t_0; t; \nu) = \Psi_\nu(t, t_0)\chi_0 \ .$$

From this, the observability grammian enables to characterize the observability of the system. Indeed, defining observability at time t_0 as the fact that for all (χ_1, χ_2),

[2]See for instance [2].

$$Y(\chi_1, t_0; t; \nu) = Y(\chi_2, t_0; t; \nu) \quad \forall t \geq t_0 \qquad \Longrightarrow \qquad \chi_1 = \chi_2 \,,$$

by linearity, this is equivalent to the fact that for all χ_0,

$$Y(\chi_0, t_0; t; \nu) = C(\nu(t))\Psi_\nu(t, t_0)\chi_0 = 0 \quad \forall t \geq t_0 \qquad \Longrightarrow \qquad \chi_0 = 0 \,,$$

which in turn is equivalent to the fact that there exists $t_1 \geq t_0$ such that $\Gamma_\nu(t_0, t_1)$ is positive definite (see [2] for more details). But we will see that for observer design, in particular, the Kalman-like designs introduced in the following chapters, it is more handy to use the backward observability grammian whose invertibility rather characterizes observability in backward time (sometimes called determinability), namely that for all (χ_1, χ_2),

$$Y(\chi_1, t_1; t; \nu) = Y(\chi_2, t_1; t; \nu) \quad \forall t \leq t_1 \qquad \Longrightarrow \qquad \chi_1 = \chi_2 \,.$$

References

1. Bonnabel, S., Slotine, J.J.: A contraction theory-based analysis of the stability of the deterministic extended Kalman filter. IEEE Trans. Autom. Control **60**(2), 565–569 (2015)
2. Chen, C.T.: Linear System Theory and Design. CBS College Publishing, New York (1984)
3. Geld, A.: Applied Optimal Estimation. MIT Press, Cambridge (1974)

Chapter 3
State-Affine Normal Forms

In this chapter, we consider systems with dynamics of the form

$$\dot{\xi} = A(u, y)\,\xi + B(u, y) \quad , \quad y = H(\xi, u) \tag{3.1}$$

where ξ is a vector of $\mathbb{R}^{d_\xi}$, and $A : \mathbb{R}^{d_u} \times \mathbb{R}^{d_y} \to \mathbb{R}^{d_\xi \times d_\xi}$, $B : \mathbb{R}^{d_u} \times \mathbb{R}^{d_y} \to \mathbb{R}^{d_\xi}$, and $H : \mathbb{R}^{d_u} \times \mathbb{R}^{d_y} \to \mathbb{R}^{d_y}$ are continuous functions.

3.1 Constant Linear Part: Luenberger Design

In this section, we consider the case where A is constant, with two subcases:

- A is Hurwitz and H any continuous function
- A is any matrix, but H is linear.

3.1.1 A Hurwitz: Luenberger's Original Form

We introduce the following definition:

Definition 3.1 We call *Hurwitz form* dynamics of the type:

$$\dot{\xi} = A\,\xi + B(u, y) \quad , \quad y = H(\xi, u)\ . \tag{3.2}$$

where A is a Hurwitz matrix in $\mathbb{R}^{d_\xi \times d_\xi}$ and B and H are continuous functions.

For a Hurwitz form, a trivial observer is made of a copy of the dynamics of the system:

P. Bernard, *Observer Design for Nonlinear Systems*,
Lecture Notes in Control and Information Sciences 479,
https://doi.org/10.1007/978-3-030-11146-5_3

Theorem 3.1 *The system*

$$\dot{\hat{\xi}} = A\hat{\xi} + B(u, y) \tag{3.3}$$

is an observer for System (3.2).

Proof The error $\hat{\xi} - \xi$ decays exponentially according to dynamics $\dot{\overbrace{\hat{\xi} - \xi}} = A(\hat{\xi} - \xi)$. □

We have referred to this form as "Luenberger's original form" because originally in [7], Luenberger's methodology to build observers for linear systems was to look for an invertible transformation which would map the linear system into a Hurwitz one, which admits a very simple observer. We will study in Part III under which condition a standard nonlinear system can be transformed into such a form, namely extending Luenberger's methodology to nonlinear systems.

3.1.2 *H Linear: $H(\xi, u) = C\xi$ with C Constant*

We consider now a system of the form[1]

$$\dot{\xi} = A\xi + B(u, y) \quad , \quad y = C\xi \tag{3.4}$$

where B is a continuous function. The following well-known result can be deduced from [7].

Theorem 3.2 *If the pair (A, C) is detectable,*[2] *there exists a matrix K such that $A - KC$ is Hurwitz. For any such matrix K, the system*

$$\dot{\hat{\xi}} = A\hat{\xi} + B(u, y) + K(y - C\hat{\xi}) \tag{3.5}$$

is an observer for System (3.4).

As opposed to Theorem 3.1, A is not supposed Hurwitz but H is a linear function.

3.2 Time-Varying Linear Part: Kalman Design

We suppose in this section that H is linear, but not necessarily constant namely

$$\dot{\xi} = A(u, y)\xi + B(u, y) \quad , \quad y = C(u)\xi \ . \tag{3.6}$$

[1] In [1], the authors propose an observer for a more general form $\dot{\xi} = A\xi + B(u, y) + G\rho(H\xi)$, $y = C\xi$, under certain conditions on ρ.

[2] If (A, C) is observable, the eigenvalues of $A - KC$ can be chosen arbitrarily.

The most famous observer used for this kind of system is the Kalman and Bucy's observer presented in [6] for linear time-varying systems, i.e., with $A(t)$, $B(t)$, and $C(t)$ replacing $A(u, y)$, $B(u, y)$, and $C(u)$, respectively. Later, a "Kalman-like" design was proposed in [2–4] for the case where $A(u, y) = A(u)$. This design can be easily extended to System (3.6) by considering (u, y) as an extended input. The difference with the time-varying case studied by Kalman and Bucy in [6] is that every assumption must be verified uniformly for any such extended input, namely for any input u and for any output function y coming from any initial condition. To highlight this fact more rigorously, we denote

$$y_{\xi_0,u}(t) = C(u(t))\,\Xi(\xi_0; t; u)$$

the output at time t of System (3.6) initialized at ξ_0 at time 0 and with input u.

Theorem 3.3 ([2–4]) *Assume the input u is such that*

- *For any ξ_0, $t \mapsto A(u(t), y_{\xi_0,u}(t))$ is bounded by A_{max}.*
- *For any ξ_0, the extended input $\nu = (u, y_{\xi_0,u})$ is regularly persistent for the auxiliary dynamics*

$$\dot{\chi} = A(u, y_{\xi_0,u})\chi \quad , \quad y = C(u)\chi \tag{3.7}$$

uniformly with respect to ξ_0; i.e., there exist strictly positive numbers t_0, $\bar{t}$, and α such that for any ξ_0 and any time $t \geq t_0 \geq \bar{t}$,

$$\Gamma_\nu^b(t - \bar{t}, t) \geq \alpha\, I$$

where Γ_ν^b is the backward observability grammian[3] *associated to System* (3.7).

Then, for any positive definite matrix P_0, there exist positive scalars α_1 and α_2 such that for any $\lambda \geq 2A_{max}$ and any $\xi_0 \in \mathbb{R}^{d_\xi}$, the matrix differential equation

$$\dot{P} = -\lambda P - A(u, y)^\top P - PA(u, y) + C(u)^\top C(u) \tag{3.8}$$

initialized at $P(0) = P_0$ admits a unique solution verifying $P(t)^\top = P(t)$ for all t, and for all $t \geq t_0$,

$$\alpha_1 I \leq P(t) \leq \alpha_2 I \ , \tag{3.9}$$

and the system

$$\dot{\hat{\xi}} = A(u, y)\,\hat{\xi} + B(u, y) + K\left(y - C(u)\hat{\xi}\right) \tag{3.10}$$

with the gain

$$K = P^{-1}C(u)^\top \tag{3.11}$$

is an observer for the state-affine system (3.6).

[3] See Definition 2.1.

Proof Let us start by studying the solutions to (3.8). Take any ξ_0 in $\mathbb{R}^{d_\xi}$ and denote $\nu = (u, y_{\xi_0,u})$. The solutions to (3.8) are given by

$$P(t) = e^{-\lambda t}\Psi_\nu^\top(0,t)P(0)\Psi_\nu(0,t) + \int_0^t e^{-\lambda(t-s)}\Psi_\nu^\top(\tau,t)C(u(\tau))^\top C(u(\tau))\Psi_\nu(\tau,t)\,d\tau\ ,$$

where Ψ_ν is the transition matrix[4] associated to the System (3.7). Therefore, $P(t)^\top = P(t)$, and because $P(0) > 0$, for all $t \geq t_0$,

$$\begin{aligned}
P(t) &\geq \int_{t-\bar{t}}^t e^{-\lambda(t-s)}\Psi_\nu^\top(\tau,t)C(u(\tau))^\top C(u(\tau))\Psi_\nu(\tau,t)d\tau \\
&\geq e^{-\lambda\bar{t}} \int_{t-\bar{t}}^t \Psi_\nu^\top(\tau,t)C(u(\tau))^\top C(u(\tau))\Psi_\nu(\tau,t)d\tau \\
&\geq e^{-\lambda\bar{t}}\,\Gamma_\nu^b\left(t-\bar{t},t\right) \\
&\geq \alpha e^{-\lambda\bar{t}}\,I
\end{aligned}$$

On the other hand, we have

$$\Psi_\nu(\tau,t) = I + \int_\tau^t A(\nu(s))\Psi_\nu(s,t)ds$$

so that for all $\tau \leq t$,

$$|\Psi_\nu(\tau,t)| = 1 + \int_\tau^t A_{max}|\Psi_\nu(s,t)|ds$$

and by Gronwall's lemma,

$$|\Psi_\nu(\tau,t)| = e^{A_{max}(t-\tau)}\ .$$

It follows from the expression of P that

$$\begin{aligned}
P(t) &\leq e^{-(\lambda-2A_{max})t}|P(0)| + \int_0^t e^{-(\lambda-2A_{max})(t-\tau)}|C(u(\tau))^\top C(u(\tau))|d\tau \\
&\leq |P(0)| + \frac{C_{max}^2}{\lambda-2A_{max}}
\end{aligned}$$

for $\lambda > 2A_{max}$. We conclude that for any positive definite matrix P_0, there exist positive scalars α_1 and α_2 such that for any ξ_0 in $\mathbb{R}^{d_\xi}$, and for any $\lambda > 2A_{max}$, the solution to (3.8) initialized at P_0 verifies (3.9).

[4]See Definition 2.1.

Consider now a positive definite matrix P_0 giving (3.9) for any ξ_0 in $\mathbb{R}^{d_\xi}$ and any $\lambda > 2A_{max}$. Consider initial conditions ξ_0 and $\hat{\xi}_0$ in $\mathbb{R}^{d_\xi}$ for Systems (3.1) and (3.10), respectively, and the corresponding solution $t \mapsto P(t)$ to (3.8). Define the function

$$V(\xi, \hat{\xi}, t) = (\hat{\xi} - \xi)^\top P(t)(\hat{\xi} - \xi) .$$

Thanks to (3.9), it verifies for all $(\xi, \hat{\xi})$ in $\mathbb{R}^{d_\xi} \times \mathbb{R}^{d_\xi}$, and all $t \geq t_0$,

$$\alpha_1 |\hat{\xi} - \xi|^2 \leq V(\xi, \hat{\xi}, t) \leq \alpha_2 |\hat{\xi} - \xi|^2 . \tag{3.12}$$

Also, according to (3.1) and (3.10),

$$\dot{\overbrace{\hat{\xi} - \xi}} = \left(A(u, y) - P^{-1} C(u)^\top C(u)\right)(\hat{\xi} - \xi)$$

so that (omitting the dependence on t to ease the notations)

$$\begin{aligned}
\dot{\overbrace{V(\xi, \hat{\xi}, t)}} &= (\hat{\xi} - \xi)^\top \left[\dot{P} + PA(u, y) - C(u)^\top C(u) + A(u, y)\top P - C(u)^\top C(u)\right](\hat{\xi} - \xi) \\
&= -\lambda\, V(\xi, \hat{\xi}, t) - (\hat{\xi} - \xi)^\top C(u)^\top C(u)(\hat{\xi} - \xi) \\
&\leq -\lambda\, V(\xi, \hat{\xi}, t) .
\end{aligned}$$

It follows that for all $t \geq t_0$,

$$V(\xi, \hat{\xi}, t) \leq e^{-\lambda(t-t_0)}\, V\left(\xi(t_0), \hat{\xi}(t_0), t_0\right)$$

which gives the result with (3.12). □

Remark 3.1

- It is important to note that K is time-varying and depends on the functions $t \mapsto u(t)$ and $t \mapsto y_{\xi_0,u}(t)$ and thus on ξ_0.
- The assumptions of boundedness of A and regular persistence are mainly to ensure that the solution to (3.8) is uniformly bounded from below and above, namely that P (and thus the gain K) goes neither to 0 nor to infinity.
- An equivalent way of writing (3.8) and (3.11) is with the Riccati equation

$$\begin{aligned}
\dot{P} &= A(u, y)P + PA(u, y)^\top - PC(u)^\top R^{-1} C(u)P + \lambda P \\
K &= P\, C(u)^\top R^{-1}
\end{aligned}$$

  with R an extra positive definite matrix to be chosen (i.e., P is replaced by P^{-1}). This implementation does not require the computation of the inverse of $P(t)$ at each step and was originally proposed in [3] as an analogy to the Kalman filter. In the case where $B(u, y) = 0$, its derivation comes from the minimization at all times of the criterion $\xi_0 \mapsto J(\xi_0, t)$ with

$$J(\xi_0, t) = e^{-\lambda t}|\xi_0 - \hat{\xi}_0|_{P_0^{-1}} + \int_0^t e^{-\lambda(t-\tau)}|y(\tau) - C(u(\tau))\Psi_\nu(\tau, 0)\xi_0|_{R^{-1}}\, d\tau$$

and by taking

$$\hat{\xi}(t) = \Psi_\nu(\tau, 0)\xi_0 ,$$

where Ψ_ν is the transition matrix[5] associated to the system (3.7). In other words, $P(0)$ represents the confidence we have in the initial condition and ξ_0 is the "best" guess of the plant's initial condition given the outputs up to time t.

- Following Kalman and Bucy's original paper [6], the gain K can also be computed with

$$\begin{aligned} \dot{P} &= A(u, y)P + PA(u, y)^\top - PC(u)^\top R^{-1}C(u)P + DQD^\top \\ K &= PC(u)^\top R^{-1} \end{aligned}$$

where R (resp Q) is a positive definite matrix representing the covariance at time t of the noise which enters the measurement (resp the dynamics) and D describes how the noise enters the dynamics. In the case where those noises are independent white noise processes, this observer solves an infinite-dimensional optimal problem: Find the optimal correction term in the observer, such that, given the values of u and y up to time t, the estimate $\hat{\xi}(t)$ of $\xi(t)$ minimizes the conditional expectation $\mathsf{E}\left(|\hat{\xi}(t) - \xi(t)|^2 \mid y_{[t_0,t]}, u_{[t_0,t]}\right)$. In order to ensure asymptotic convergence of the observer, according to [6, Theorem 4], the following assumptions are needed:

  - Boundedness of A
  - Uniform complete observability of (A, C): This corresponds to the regular persistence condition of Theorem 3.3 when A and C depend on an input u and A is bounded (see [5])
  - Uniform complete controllability of (A, D): This is the dual of uniform complete observability, namely uniform complete observability of $(A^\top, D^\top)$ (see [5])
  - R and Q are uniformly lower- and upper-bounded in time.

Only the first two assumptions depend on the system, and they are the same as in Theorem 3.3; the other two must be satisfied by an appropriate choice of the design parameters R and Q.

References

1. Arcak, M., Kokotovic, P.: Observer-based control systems with slope-restricted nonlinearities. IEEE Trans. Autom. Control **46** (2001)
2. Besançon, G., Bornard, G., Hammouri, H.: Observer synthesis for a class of nonlinear control systems. Eur. J. Control **3**(1), 176–193 (1996)

[5] See Definition 2.1.

3. Bornard, G., Couenne, N., Celle, F.: Regularly persistent observers for bilinear systems. In: Descusse, J., Fliess, M., Isidori, A., Leborgne, D. (eds.) New Trends in Nonlinear Control Theory, pp. 130–140. Springer, Berlin (1989)
4. Hammouri, H., Morales, J.D.L.: Observer synthesis for state-affine systems. In: IEEE Conference on Decision and Control, pp. 784–785 (1990)
5. Kalman, R.: Contributions to the theory of optimal control. In: Conference on Ordinary Differential Equations (1960)
6. Kalman, R., Bucy, R.: New results in linear filtering and prediction theory. J. Basic Eng. **108**, 83–95 (1961)
7. Luenberger, D.: Observing the state of a linear system. IEEE Trans. Mil. Electron. **8**, 74–80 (1964)

Chapter 4
Triangular Forms

Triangular forms became of interest when [13] related their structure to uniformly observable systems, and when [30] introduced the phase-variable form for differentially observable systems. The celebrated high-gain observer proposed in [11, 28] for phase-variable forms and later in [9, 14] for triangular forms have been extensively studied ever since. This high-gain design, originally built under a Lipschitzness condition on the triangular nonlinearities, was then extended to the larger class of homogeneous triangular forms [2, 23] and more recently to only continuous triangular forms [6]. We will see later in Part II how those latter forms are of interest for systems which are uniformly observable[1] and differentially observable at an order which is greater than the dimension of the system: In this case, the system may be transformed in a triangular form but with continuous nonlinearities which may not be locally Lipschitz.

All those high-gain designs are recalled in this chapter. In order to avoid lengthy unpleasant ruptures in the text, some proofs are only summed up, and their full version is given in Appendix B.

Notations

For $(\xi_1, \ldots, \xi_q)$ and $(\hat{\xi}_1, \ldots, \hat{\xi}_q)$ (resp. $(\hat{\xi}_{i1}, \ldots, \hat{\xi}_{iq})$) in $\mathbb{R}^q$, we denote

$$\begin{array}{c} \boldsymbol{\xi}_i = (\xi_1, \ldots, \xi_i) \quad , \quad \hat{\boldsymbol{\xi}}_i = (\hat{\xi}_1, \ldots, \hat{\xi}_i) \quad \text{(resp. } \hat{\boldsymbol{\xi}}_i = (\hat{\xi}_{i1}, \ldots, \hat{\xi}_{ii})) \\ e_{ij} = \hat{\xi}_{ij} - \xi_j \ , \quad e_j = \hat{\xi}_j - \xi_j \ , \ \boldsymbol{e}_i = \hat{\boldsymbol{\xi}}_i - \boldsymbol{\xi}_i \ . \end{array} \tag{4.1}$$

4.1 Nominal Triangular Form: High-Gain Designs

Definition 4.1 We call *continuous triangular form* dynamics of the form:

[1] See Definition 1.2.

P. Bernard, *Observer Design for Nonlinear Systems*,
Lecture Notes in Control and Information Sciences 479,
https://doi.org/10.1007/978-3-030-11146-5_4

$$\begin{cases} \dot{\xi}_1 = \xi_2 + \Phi_1(u, \xi_1) \\ \quad \vdots \\ \dot{\xi}_i = \xi_{i+1} + \Phi_i(u, \xi_1, \dots, \xi_i) \\ \quad \vdots \\ \dot{\xi}_m = \Phi_m(u, \xi) \end{cases} \quad , \quad y = \xi_1 \tag{4.2}$$

where for all i in $\{1, \dots, m\}$, ξ_i is in $\mathbb{R}^{d_y}$, $\xi = (\xi_1, \dots, \xi_m)$ is in $\mathbb{R}^{d_\xi}$, with $d_\xi = md_y$, $\Phi_i : \mathbb{R}^{d_u} \times \mathbb{R}^{id_y} \to \mathbb{R}^{d_y}$ are continuous functions. In the particular case where only Φ_m is nonzero, we say *continuous phase-variable form*.

If now the functions $\Phi_i(u, \cdot)$ are globally Lipschitz on $\mathbb{R}^{id_y}$ uniformly in u, namely there exists $\mathfrak{a}$ in $\mathbb{R}$ such that for all u in U, all (ξ_a, ξ_b) in $\mathbb{R}^{d_\xi} \times \mathbb{R}^{d_\xi}$ and for all i in $\{1, \dots, m\}$

$$|\Phi_i(u, \xi_{1a}, \dots, \xi_{ia}) - \Phi_i(u, \xi_{1b}, \dots, \xi_{ib})| \leq \mathfrak{a} \sum_{j=1}^{i} |\xi_{ja} - \xi_{jb}| ,$$

we say *Lipschitz triangular form* and *Lipschitz phase-variable form*.

Actually, we will see in this section that the regularity of the functions Φ_i conditions the convergence of certain observers. Wanting to present the results in a unified and concise way, we define the following property as in [6].

Definition 4.2 (*Property* $\mathscr{H}(\alpha, \mathfrak{a})$) For a positive real number $\mathfrak{a}$ and a vector α in $[0, 1]^{\frac{m(m+1)}{2}}$, we will say that the function Φ verifies the property[2] $\mathscr{H}(\alpha, \mathfrak{a})$ if for all i in $\{1, \dots, m\}$, for all (ξ_a, ξ_b) in $\mathbb{R}^{d_\xi} \times \mathbb{R}^{d_\xi}$ and u in U, we have:

$$|\Phi_i(u, \xi_{1a}, \dots, \xi_{ia}) - \Phi_i(u, \xi_{1b}, \dots, \xi_{ib})| \leq \mathfrak{a} \sum_{j=1}^{i} |\xi_{ja} - \xi_{jb}|^{\alpha_{ij}} . \tag{4.3}$$

This property captures many possible contexts. In the case in which $\alpha_{ij} > 0$, it implies that the function Φ is Hölder with power α_{ij} and when all $\alpha_{ij} = 1$, we recover the Lipschitz triangular form. When the $\alpha_{ij} = 0$, it simply implies that the function Φ is bounded.

4.1.1 Lipschitz Triangular Form

The Lipschitz triangular form is well known because it allows the design of a high-gain observer.

---

[2]This property can be relaxed: See Sect. 4.1.5.

Theorem 4.1 ([14]) *Assume the function Φ verifies $\mathscr{H}(\alpha, \mathfrak{a})$ for some $\mathfrak{a}$ in $\mathbb{R}_+$ and with $\alpha_{ij} = 1$ for all $1 \leq j \leq i \leq m$, namely the functions $\Phi_i(u, \cdot)$ are globally Lipschitz on $\mathbb{R}^{id_y}$, uniformly in u. For any $(k_1, \ldots, k_m)$ in $\mathbb{R}^m$ such that the roots of the polynomial*

$$s^m + k_m s^{m-1} + \cdots + k_2 s + k_1$$

have strictly negative real parts, there exist positive real numbers λ, β, γ, and L^* such that, for all $L \geq \max\{\mathfrak{a} L^*, 1\}$, for all u in $\mathscr{U}$ and all $(\xi_0, \hat{\xi}_0)$ in $\mathbb{R}^{d_\xi} \times \mathbb{R}^{d_\xi}$, any solution $\hat{\Xi}(\hat{\xi}_0; t; u, y_{\xi_0})$ of

$$\left\{ \begin{array}{rl} \dot{\hat{\xi}}_1 = & \hat{\xi}_2 + \Phi_1(u, \hat{\xi}_1) - L\, k_1\, (\hat{\xi}_1 - y) \\ \dot{\hat{\xi}}_2 = & \hat{\xi}_3 + \Phi_2(u, \hat{\xi}_1, \hat{\xi}_2) - L^2\, k_2\, (\hat{\xi}_1 - y) \\ \vdots & \\ \dot{\hat{\xi}}_m = & \Phi_m(u, \hat{\xi}) - L^m\, k_m\, (\hat{\xi}_1 - y) \end{array} \right. \tag{4.4}$$

and any solution $\Xi(\xi_0; t; u)$ of (4.2) *verify, for all t_0 and t such that $t \geq t_0 \geq 0$, and for all i in $\{1, \ldots, m\}$,*

$$\left| \hat{\Xi}_i(t) - \Xi_i(t) \right| \leq L^{i-1} \beta \left| \hat{\Xi}(t_0) - \Xi(t_0) \right| e^{-\lambda L (t - t_0)}$$

where we have used the abbreviation $\Xi(t) = \Xi(\xi_0; t; u)$ and $\hat{\Xi}(t) = \hat{\Xi}(\hat{\xi}_0; t; u, y_{\xi_0})$. In other words, (4.4) *is an observer for the Lipschitz triangular form* (4.2).

Proof We give here the general idea of the proof, and its full version is available in Appendix B.1.1. The error $e = \hat{\xi} - \xi$ produced by observer (4.4) satisfies

$$\dot{e} = LAe + \Phi(u, \hat{\xi}) - \Phi(u, \xi) - L\mathscr{L}KCe$$

where

$$A = \begin{pmatrix} 0 & I_{d_y} & 0 & \cdots & 0 \\ \vdots & \ddots & \ddots & & \vdots \\ \vdots & & \ddots & \ddots & 0 \\ & & & 0 & I_{d_y} \\ 0 & \cdots & \cdots & 0 & 0 \end{pmatrix} , \quad \mathscr{L} = \begin{pmatrix} I_{d_y} & & & \\ & L I_{d_y} & & \\ & & \ddots & \\ & & & L^{m-1} I_{d_y} \end{pmatrix}$$

$$C = \begin{pmatrix} I_{d_y} & 0 \cdots 0 \end{pmatrix} , \quad K = \begin{pmatrix} k_1 I_{d_y} \\ k_2 I_{d_y} \\ \vdots \\ k_m I_{d_y} \end{pmatrix} .$$

The main idea of the high-gain designs is to consider the scaled error coordinates $\varepsilon = \mathscr{L}^{-1} e$ where the error dynamics read

$$\frac{1}{L}\dot{\varepsilon} = (A - KC)\varepsilon + \frac{1}{L}\mathscr{L}^{-1}\left(\Phi(u,\hat{\xi}) - \Phi(u,\xi)\right) .$$

Since the matrix $A - KC$ is Hurwitz, we have an asymptotically stable system disturbed by $\mathscr{L}^{-1}\left(\Phi(u,\hat{\xi}) - \Phi(u,\xi)\right)$, which vanishes when $\varepsilon = 0$ and whose size depends on the Lipschitz constant of Φ. Actually, thanks to the triangularity of Φ, it is possible to show that $\mathscr{L}^{-1}\left(\Phi(u,\hat{\xi}) - \Phi(u,\xi)\right)$ is bounded independently from L by "$\mathfrak{a}\varepsilon$". Therefore, thanks to the factor $\frac{1}{L}$, by taking L sufficiently large, its impact on the system can be made arbitrarily small and by robustness of the Lyapunov function associated to $A - KC$, convergence is ensured. Note that due to the factor $\frac{1}{L}$ in front of $\dot{\varepsilon}$, increasing the gain has also the effect of accelerating time, hence a convergence rate linear in L.

Observer (4.4) is called a high-gain observer because the gain L must be chosen sufficiently large in order to compensate for the Lipschitz constant of the nonlinearities, which are seen here as disturbances. Notice that this observer provides an exponential convergence whose rate increases with L. However, the larger L, the larger the solution can become before converging due to the factor L^{i-1} in front of the error bound: This is the so-called peaking phenomenon. It can also be shown that the larger L, the larger the impact of disturbances and measurement noise, so that a compromise has to be found when choosing L. We refer the interested reader to [17] and the references therein for a complete analysis of this high-gain design.

Actually, extensions of this high-gain observer exist for more complex triangular forms, in particular when each block does not have the same dimension, but extra assumptions on the dependence of the function Φ_i must be made to ensure convergence (see [9] or later [16] for instance).

In any case, the standard implementation of a high-gain observer necessitates the global Lipschitzness of the nonlinearities Φ_i. If they are only locally Lipschitz, it is still possible to use observer (4.4) if the trajectories of the system evolve in a compact set, by saturating Φ_i outside this compact set (see Sect. 4.1.5). Otherwise, it has been tried to adapt the high-gain L online by "following" the Lipschitz constant of Φ_i when it is observable from the output ([3, 4, 22, 26] and references therein).

4.1.2 High-Gain Observer For a Non-Lipschitz Triangular Form?

Unfortunately, it is shown in [6] that when the nonlinearities are non-Lipschitz, the asymptotic convergence of the high-gain observer can be lost. Nevertheless, it has been known for a long time, mostly in the context of dirty-derivatives and output differentiation, that in the particular case of a phase-variable form, a high-gain observer can provide an arbitrary small error when Φ_m is only bounded ([28] among many others). This result was extended in [6] to full triangular forms under some Hölder-like conditions given in Table 4.1.

Table 4.1 Hölder restrictions (4.5) on Φ for arbitrarily small errors with a high-gain observer: the Hölder powers α_{ij} in (4.3) must be greater than the values entered in the table for i = 1...m-1, and greater or equal for $i = m$

i \ j	1	2	...	m-2	m-1	m
1	$\frac{m-2}{m-1}$					
2	$\frac{m-3}{m-2}$	$\frac{m-3}{m-2}$				
⋮	⋮	⋮	⋱			
m-2	$\frac{1}{2}$	$\frac{1}{2}$	...	$\frac{1}{2}$		
m-1	0	0	...	...	0	
m	0	0	...	...	...	0

Theorem 4.2 ([6]) *Assume the function Φ verifies $\mathscr{H}(\alpha, \mathfrak{a})$ for some $(\alpha, \mathfrak{a})$ in $[0, 1]^{\frac{m(m+1)}{2}} \times \mathbb{R}_+$ satisfying, for $1 \leq j \leq i$*

$$\begin{aligned} \tfrac{m-i-1}{m-i} < \alpha_{ij} \leq 1 \quad \textit{for} \quad i = 1 \ldots, m-1 \,, \\ 0 \leq \alpha_{mj} \leq 1 \end{aligned} \tag{4.5}$$

as shown in Table 4.1. Then, for any $\varepsilon > 0$ and for any $(k_1, \ldots, k_m)$ in $\mathbb{R}^m$ such that the roots of the polynomial

$$s^m + k_m s^{m-1} + \cdots + k_2 s + k_1$$

have strictly negative real parts, there exist positive real numbers λ, β, γ, and L^* such that, for all $L \geq L^*$, for all u in $\mathscr{U}$ and all $(\xi_0, \hat{\xi}_0)$ in $\mathbb{R}^{d_\xi} \times \mathbb{R}^{d_\xi}$, any solution $\hat{\Xi}(\hat{\xi}_0; t; u, y_{\xi_0})$ of (4.4) verifies, for all t_0 and t such that $t \geq t_0 \geq 0$, and for all i in $\{1, \ldots, m\}$,

$$\left|\hat{\Xi}_i(t) - \Xi_i(t)\right| \leq \max\left\{\varepsilon \,,\; L^{i-1}\beta \left|\hat{\Xi}(t_0) - \Xi(t_0)\right| e^{-\lambda L(t-t_0)}\right\}$$

where we have used the abbreviation $\Xi(t) = \Xi(\xi_0; t; u)$ and $\hat{\Xi}(t) = \hat{\Xi}(\hat{\xi}_0; t; u, y_{\xi_0})$.

Proof See Appendix B.1.2.

This result says that under a Hölder condition instead of the Lipschitz one, the asymptotic convergence of the high-gain observer may be lost, but at least a "practical convergence" is maintained, namely it is possible to ensure arbitrarily small errors with a sufficiently large gain.

It is interesting to remark the weakness of the assumptions imposed on the last two components of the function Φ. Indeed, (4.5) only imposes that $\Phi_{d_\xi-1}$ be Hölder without any restriction on the order, and that Φ_{d_ξ} be bounded.[3]

In the next section, we show that under slightly stronger Hölder constraints, asymptotic convergence can actually be achieved when considering homogeneous observers.

4.1.3 Hölder Continuous Triangular Form

Fortunately, moving to a generalization of high-gain observers exploiting homogeneity makes it possible to achieve convergence in the case of non-Lipschitz nonlinearities verifying some Hölder conditions. It is at the beginning of the century that researchers started to consider homogeneous observers with various motivations: exact differentiators [18–20], domination as a tool for designing stabilizing output feedback [2, 23, 25, 29], and references therein (in particular [1]). As shown in [6], the advantage of this type of observers is their ability to face Hölder nonlinearities.

Theorem 4.3 ([6, 23]) *Assume that there exist d_0 in $[-1, 0]$ and $\mathfrak{a}$ in $\mathbb{R}_+$ such that Φ satisfies $\mathscr{H}(\alpha, \mathfrak{a})$ with α verifying*[4]*:*

$$\alpha_{ij} = \frac{1 - d_0(m-i-1)}{1 - d_0(m-j)} = \frac{r_{i+1}}{r_j} \;, \quad 1 \leq j \leq i \leq m \;, \tag{4.6}$$

where r is a vector in $\mathbb{R}^{m+1}$, called weight vector, the components of which, called weights, are defined by

$$r_i = 1 - d_0(m-i) \;. \tag{4.7}$$

There exist $(k_1, \dots, k_m)$ in $\mathbb{R}^m$ and $L^* \geq 1$ such that, for all $L \geq L^*$, for any u in $\mathscr{U}$, for any initial conditions $(\xi_0, \hat{\xi}_0)$ in $\mathbb{R}^{d_\xi} \times \mathbb{R}^{d_\xi}$, the system

$$\left\{\begin{array}{l} \dot{\hat{\xi}}_1 = \hat{\xi}_2 + \Phi_1(u, \hat{\xi}_1) - L\, k_1 \left\lfloor \hat{\xi}_1 - y \right\rceil^{\frac{r_2}{r_1}} \\ \dot{\hat{\xi}}_2 = \hat{\xi}_3 + \Phi_2(u, \hat{\xi}_1, \hat{\xi}_2) - L^2\, k_2 \left\lfloor \hat{\xi}_1 - y \right\rceil^{\frac{r_3}{r_1}} \\ \quad\vdots \\ \dot{\hat{\xi}}_m \in \Phi_m(u, \hat{\xi}) - L^m\, k_m \left\lfloor \hat{\xi}_1 - y \right\rceil^{\frac{r_{m+1}}{r_1}} \end{array}\right. \tag{4.8}$$

[3] See Sect. 4.1.5.

[4] This may be relaxed: See Sect. 4.1.5.

with the notation[5]

$$\lfloor a \rceil^b = \mathtt{sign}(a)\,|a|^b \quad , \quad b > 0$$

and

$$\lfloor a \rceil^0 = \mathcal{S}(a) = \begin{cases} \{1\} & \text{if } a > 0\,, \\ [-1, 1] & \text{if } a = 0\,, \\ \{-1\} & \text{if } a < 0\,, \end{cases} \tag{4.9}$$

admits absolutely continuous solutions defined on $\mathbb{R}_+$, *and is an observer for the continuous triangular form* (4.2). *Moreover, when* $d_0 < 0$, *this observer converges in finite time.*

Proof The proof of convergence relies on

1. recursively building an homogeneous Lyapunov function in the high-gain error coordinates for the system without any nonlinearity
2. thanks to the robustness of the Lyapunov function, showing that the nonlinearities can be dominated for a sufficient large gain.

See Appendix B.2. □

d_0 is called degree of the observer. When $d_0 = 0$, all the weights r_i are equal to 1, the nonlinearities are Lipschitz, and we recover the high-gain observer (4.4). In that sense, we can say that the homogeneous observer (4.8) is an extension of (4.4). Its convergence is proved in [23] for $d_0 \in (-1, 0]$.

In the limit case where $d_0 = -1$, r_{m+1} vanishes, which makes the last correction term of (4.8) equal to $\left\lfloor \hat{\xi}_1 - y \right\rceil^0 = \mathtt{sign}(\hat{\xi}_1 - y)$. This function being discontinuous at 0, the system becomes a differential inclusion[6] when defining the sign function as the set-valued map (4.9). Note that this set-valued map is upper semi-continuous with nonempty, compact, and convex values, namely it verifies the usual basic conditions for existence of absolutely continuous solutions for differential inclusions given in [12, 27].

Actually, when $d_0 = -1$, we recover the same correction terms as in the exact differentiator presented in [18], where finite-time convergence is established for a phase-variable form with Φ_m is bounded. Quite naturally, this boundedness condition on Φ_m is exactly the condition we obtain when taking $d_0 = -1$ in the Hölder constraint (4.6). It was then shown in [6] that this exact differentiator can also be used in presence of continuous nonlinearities on every line, provided they verify the Hölder constraint (4.6) with $d_0 = -1$ (see Table 4.2). This extreme case is of great interest because it is when the Hölder constraints are the least restrictive.

Note that a generalization of observer (4.8) was presented in [2] in the context of "bi-limit" homogeneity, i.e., for nonlinearities having two homogeneity degrees (around the origin and around infinity), namely

[5]If ξ_1 is a block of dimension d_y, those notations apply component-wise, namely $\lfloor a \rceil^b = (\lfloor a_1 \rceil^b, \ldots, \lfloor a_m \rceil^b)^\top$.

[6]See Remark 1.1.

Table 4.2 Hölder restrictions (4.6) on Φ for homogeneous observer (4.8) with $d_0 = -1$: the Hölder powers α_{ij} in (4.3) must be greater or equal to the values entered in the table (see Sect. (4.1.5))

i \ j	1	2	...	m-2	m-1	m
1	$\frac{m-1}{m}$					
2	$\frac{m-2}{m}$	$\frac{m-2}{m-1}$				
⋮	⋮	⋮	⋱			
m-2	$\frac{2}{m}$	$\frac{2}{m-1}$	...	$\frac{2}{3}$		
m-1	$\frac{1}{m}$	$\frac{1}{m-1}$	...	...	$\frac{1}{2}$	
m	0	0	...	...	...	0

$$|\Phi_i(u, \xi_{1a}, \ldots, \xi_{ia}) - \Phi_i(u, \xi_{1b}, \ldots, \xi_{ib})| \leq \mathfrak{a}_0 \sum_{j=1}^{i} |\xi_{ja} - \xi_{jb}|^{\frac{r_{0,i+1}}{r_{0,j}}} + \mathfrak{a}_\infty \sum_{j=1}^{i} |\xi_{ja} - \xi_{jb}|^{\frac{r_{\infty,i+1}}{r_{\infty,j}}} ,$$

with

$$r_{0,i} = 1 - d_0(m-i) \quad , \quad r_{\infty,i} = 1 - d_\infty(m-i)$$

and $-1 < d_0 \leq d_\infty < \frac{1}{m+1}$.

Remark 4.1 As opposed to [18] where convergence is established only for a phase-variable form via a solution-based analysis, convergence is here guaranteed for the more general triangular form by construction of a a strict homogeneous Lyapunov function which allows the presence of homogeneous disturbances in the dynamics. Actually, many efforts have been made to get expressions of Lyapunov functions for the output differentiator from [18]. First limited to small dimensions in [21], it was only recently achieved in [10]. This approach is in fact much harder since the authors look for a Lyapunov function for an already existing observer (Lyapunov analysis), while in [2, 6, 24], the observer and the Lyapunov function are built at the same time (Lyapunov design). See Appendix B.2.

4.1.4 Continuous Triangular Form

The homogeneous observer presented in the previous subsection requires the nonlinearities to verify some Hölder conditions, the least restrictive of which are given in Table 4.2 for $d_0 = -1$. To face the unfortunate situation where the nonlinearities verify none of those Hölder-type conditions, it is interesting to build an observer that works for any continuous nonlinearities.

The first observer able to cope with Φ no more than continuous is the one presented in [5] for $d_y = 1$. Its dynamics are described by a differential inclusion[7]

$$\dot{\hat{\xi}} \in \mathscr{F}(\hat{\xi}, y, u)$$

where $(\hat{\xi}, y, u) \mapsto \mathscr{F}(\hat{\xi}, y, u)$ is a set-valued map defined by: $(v_1, \ldots, v_m)$ is in $\mathscr{F}(\hat{\xi}, y, u)$ if there exists $(\tilde{\xi}_2, \ldots, \tilde{\xi}_m)$ in $\mathbb{R}^{m-1}$ such that [8]

$$\begin{aligned}
v_1 &= \tilde{\xi}_2 + \Phi_1(u, y) \\
\tilde{\xi}_2 &\in \mathsf{sat}_{M_2}(\hat{\xi}_2) - k_1 \, \mathrm{S}(y - \hat{\xi}_1) \\
&\vdots \\
v_i &= \tilde{\xi}_{i+1} + \Phi_i(u, y, \tilde{\xi}_2, \ldots, \tilde{\xi}_i) \\
\tilde{\xi}_{i+1} &\in \mathsf{sat}_{M_{i+1}}(\hat{\xi}_{i+1}) - k_i \, \mathrm{S}(\hat{\xi}_i - \tilde{\xi}_i) \\
&\vdots \\
v_m &\in \Phi_m(u, y, \tilde{\xi}_2, \ldots, \tilde{\xi}_m) - k_m \, \mathrm{S}(\hat{\xi}_m - \tilde{\xi}_m)
\end{aligned}$$

where M_i are known bounds for each components of the solution. It can be shown that any absolutely continuous solution gives in finite time an estimate of ξ under the only assumption of boundedness of the input and of the state trajectory. But the set-valued map $\mathscr{F}$ above does not satisfy the usual basic assumptions given in [12, 27] (upper semi-continuous with nonempty, compact, and convex values). It follows that we are not guaranteed of existence of absolutely continuous solutions nor of possible sequential compactness of such solutions and therefore of possibilities of approximations of $\mathscr{F}$.

Another solution recently introduced in [6] consists of a cascade of homogeneous observer, based on the observation that, for α verifying (4.6) with $d_0 = -1$, $\mathscr{H}(\alpha, \mathfrak{a})$ does not impose any restriction besides boundedness of the last functions Φ_m (see Table 4.2). Indeed, from the remark that observer (4.8)

1. can be used for the system

$$\begin{aligned}
\dot{\xi}_1 &= \xi_2 + \psi_1(t) \\
&\vdots \\
\dot{\xi}_{k-1} &= \xi_k + \psi_{k-1}(t) \\
\dot{\xi}_k &= \varphi_k(t)
\end{aligned}$$

[7] See Remark 1.1.

[8] The saturation function is defined on $\mathbb{R}$ by $\mathsf{sat}_a(x) = \max\{\min\{x, a\}, -a\}$.

provided the functions ψ_i are known and the function φ_k is unknown but bounded, with known bound.

2. gives estimates of the ξ_i's in finite time,

we see that it can be used as a preliminary step to deal with the system

$$\begin{aligned}\dot{\xi}_1 &= \xi_2 + \psi_1(t)\\ &\vdots\\ \dot{\xi}_{k-1} &= \xi_k + \psi_{k-1}(t)\\ \dot{\xi}_k &= \xi_{k+1} + \Phi_k(u, \xi_1, \ldots, \xi_k)\\ \dot{\xi}_{k+1} &= \varphi_{k+1}(u, \xi_1, \ldots, \xi_{k+1})\end{aligned}$$

Indeed, thanks to the above observer we know in finite time the values of $\xi_1, \ldots, \xi_k$, so that the function $\Phi_k(u, \xi_1, \ldots, \xi_k)$ becomes a known signal $\psi_k(t)$. Therefore, as a direct consequence of Theorem 4.3, we have:

Theorem 4.4 ([6]) *Assume Φ is continuous and there exist positive real numbers $\overline{\xi}$ and $\overline{u}$, and a subset $\mathcal{M}_0$ of $\mathbb{R}^{d_\xi}$ such that for any ξ_0 in $\mathcal{M}_0$, any u in $\mathcal{U}$, and any solution $\Xi(\xi_0, t; u)$ to (4.2),*

$$|\Xi(\xi_0, t; u)| \leq \overline{\xi} \quad , \quad |u(t)| \leq \overline{u} \qquad \forall t \geq 0\,.$$

There exist positive real numbers k_{ij} and L_i^* such that, for all $(L_1, \ldots, L_m)$ verifying $L_i \geq L_i^*$, for all input u in $\mathcal{U}$, and all $(\xi_0, \hat{\xi}_0)$ in $\mathcal{M}_0 \times \mathbb{R}^{d_\xi}$, the system

$$\left\{\begin{array}{l}
\dot{\hat{\xi}}_{11} \in -L_1\, k_{11}\, S(\hat{\xi}_{11} - y)\\
\cdots\cdots\cdots\cdots\cdots\cdots\cdots\cdots\cdots\cdots\cdots\cdots\\
\dot{\hat{\xi}}_{21} = \hat{\xi}_{22} + \Phi_1(u, \hat{\xi}_{11}) - L_2\, k_{21} \left\lfloor \hat{\xi}_{21} - y \right\rceil^{\frac{1}{2}}\\
\dot{\hat{\xi}}_{22} \in -L_2^2\, k_{22}\, S(\hat{\xi}_{21} - y)\\
\cdots\cdots\cdots\cdots\cdots\cdots\cdots\cdots\cdots\cdots\cdots\cdots\\
\quad\vdots\\
\cdots\cdots\cdots\cdots\cdots\cdots\cdots\cdots\cdots\cdots\cdots\cdots\\
\dot{\hat{\xi}}_{d_\xi 1} = \hat{\xi}_{m2} + \Phi_1(u, \hat{\xi}_{11}) - L_m\, k_{m1} \left\lfloor \hat{\xi}_{m1} - y \right\rceil^{\frac{m-1}{m}}\\
\quad\vdots\\
\dot{\hat{\xi}}_{m(m-1)} = \hat{\xi}_{mm} + \Phi_{m-1}(u, \hat{\xi}_{(m-1)1}, \ldots, \hat{\xi}_{(m-1)(m-1)}) - L_m^{m-1}\, k_{m(m-1)} \left\lfloor \hat{\xi}_{m1} - y \right\rceil^{\frac{1}{m}}\\
\dot{\hat{\xi}}_{mm} \in -L_m^m\, k_{mm}\, S(\hat{\xi}_{m1} - y)
\end{array}\right. \tag{4.10}$$

admits absolutely continuous solutions $\left(\hat{\boldsymbol{\Xi}}_1(\hat{\xi}_0; t; u, y_{\xi_0}), \ldots, \hat{\boldsymbol{\Xi}}_m(\hat{\xi}_0; t; u, y_{\xi_0})\right)$ defined on $\mathbb{R}_+$, and gives a finite-time observer to the continuous triangular form (4.2), in the sense that for any such solutions, there exists $\overline{t}$ such that for all i in $\{1, \ldots, m\}$,

$$\hat{\Xi}_i(t) = \Xi_i(t) \qquad \forall t \geq \overline{t}$$

where $\hat{\Xi}_i$ *is the state of the ith block (see Notation* (4.1)*) and we have used the abbreviation* $\hat{\Xi}_i(t) = \hat{\Xi}_i(\hat{\xi}_0, \xi_0; t; u, y_{\xi_0})$ *and* $\Xi_i(t) = \Xi_i(\xi_0; t; u)$.

Proof The proof relies on the fact that the error system is a cascade of ISS error systems which successively converge to 0 in finite time. It can be found in [6].

The use of an homogeneous cascade enables here to obtain asymptotic (even finite-time) convergence without demanding anything but continuity of the nonlinearities, and boundedness of the input and of the system solutions. A drawback of this cascade of observers is that it gives an observer with dimension $\frac{m(m+1)}{2}$ in general. However, it may be possible to reduce this dimension since, for each new block, one may increase the dimension by more than one, when the corresponding added functions Φ_i satisfy $\mathscr{H}(\alpha, \mathfrak{a})$ for some α verifying (4.6) with $d_0 = -1$ and for some $\mathfrak{a}$.

Remark 4.2 A similar cascade could be built with standard high-gain observers instead of homogeneous observers. Indeed, the conditions (4.5) for practical convergence given in Theorem 4.2 do not ask for anything but boundedness of the last nonlinearity (see Table 4.1). It is proved in [6] that still assuming boundedness of input and system trajectories, such a cascade provides practical convergence for any continuous triangular form. In other words, the final error can be made arbitrarily small by choosing the gains of each block appropriately. However, this choice of gain is more delicate and only practical convergence is ensured.

4.1.5 Relaxation of Some Assumptions

The global aspect of boundedness, Hölder, $\mathscr{H}(\alpha, \mathfrak{a})$, ..., can be relaxed[9] as follows. Let U be a bounded subset of $\mathbb{R}^{d_u}$ and let $\mathscr{M}$ be a compact subset of $\mathbb{R}^{d_\xi}$. We define $\hat{\Phi}$ as

$$\hat{\Phi}_i(u, \xi_1, \ldots, \xi_i) = \mathsf{sat}_{\overline{\Phi}_i}(\Phi_i(u, \xi_1, \ldots, \xi_i)) \tag{4.11}$$

where

$$\overline{\Phi}_i = \max_{u \in U, \xi \in \mathscr{M}} |\Phi_i(u, \xi_1, \ldots, \xi_i)| \,.$$

Now consider any compact set $\tilde{\mathscr{M}}$ strictly contained[10] in $\mathscr{M}$. We have $\hat{\Phi} = \Phi$ on $\tilde{\mathscr{M}}$, so that if the system trajectories remain in $\tilde{\mathscr{M}}$, the model made of the triangular form (4.2) with $\hat{\Phi}$ replacing Φ is still valid. Besides, according to Lemma A.7 in Appendix A.2, there exists $\tilde{\mathfrak{a}}$ such that (4.3) holds for $\hat{\Phi}$ for all $(\boldsymbol{\xi}_a, \boldsymbol{\xi}_b)$ in $\mathbb{R}^{d_\xi} \times \tilde{\mathscr{M}}$. Then, by taking $\hat{\Phi}$ instead of Φ in the observers, we can modify the assumptions in

[9] Section 4.1.5 is reproduced from [6] with permission from Elsevier.

[10] By strictly contained, we mean that $\tilde{\mathscr{M}}$ is contained in the interior of $\mathscr{M}$.

Theorems 4.1, 4.2, and 4.3, so that Φ verifies $\mathcal{H}(\alpha, \mathfrak{a})$ only on the compact set $\mathcal{M}$. In this case, the results hold for the system solutions $\Xi(\xi_0; t; u)$ which are in the compact set $\tilde{\mathcal{M}}$ for all t. Otherwise, for these solutions, the bounds on $\hat{\Xi}_i(t) - \Xi_i(t)$ obtained in these theorems hold for all t in $[0, \sigma^+_{\tilde{\mathcal{M}}}(\xi_0, u))$.

Note also that if $\mathcal{H}(\alpha, \mathfrak{a})$ holds on a compact set, then for any $\tilde{\alpha}$ such that $\tilde{\alpha}_{ij} \leq \alpha_{ij}$ for all (i, j), there exists $\tilde{\mathfrak{a}}$ such that $\mathcal{H}(\tilde{\alpha}, \tilde{\mathfrak{a}})$ also holds on this compact set. It follows that the constraints given by (4.6) in Theorem 4.3 can be relaxed to $\alpha_{ij} \geq \frac{1-d_0(m-i-1)}{1-d_0(m-j)}$.

Example 4.1 As an example, we consider the triangular normal form of dimension 4 defined by

$$\begin{cases} \dot{\xi}_1 = \xi_2 \\ \dot{\xi}_2 = \xi_3 \\ \dot{\xi}_3 = \xi_4 + \Phi_3(u, \xi_1, \xi_2, \xi_3) \\ \dot{\xi}_4 = \Phi_4(u, \xi) \end{cases} , \quad y = \xi_1 , \tag{4.12}$$

where

$$\Phi_3(u, \xi_1, \xi_2, \xi_3) = 5u|\xi_3 + \xi_1|^{\frac{4}{5}} \lfloor \xi_1 \rceil^{\frac{1}{5}} \quad , \quad \Phi_4(u, \xi) = \Psi(u, \psi(\xi))$$

with $\Psi : \mathbb{R}^{d_u} \times \mathbb{R}^3 \to \mathbb{R}$ and $\psi : \mathbb{R}^4 \to \mathbb{R}^3$ continuous function defined by

$$\begin{aligned} \Psi(u, \psi) = \psi_1 - 2\psi_1\psi_3^5 + 20\psi_3^3\psi_1^3\psi_2^2 - 15\psi_3^4\psi_2^2\psi_1 \\ + 5\psi_3^4\psi_1^3 - 5\psi_3^9\psi_1^3 + \psi_3^{10}\psi_1 + u(-20\psi_3^3\psi_1^2\psi_2 + 5\psi_3^4\psi_2) \end{aligned}$$

$$\psi(\xi) = \left(\xi_1 \ , \ \xi_2 \ , \ \left(\frac{(\xi_3 + \xi_1)\xi_1 + \left[(\xi_4 + \xi_2) + 3|(\xi_3 + \xi_1) \lfloor \xi_1 \rceil^{\frac{3}{2}} |^{\frac{4}{5}} \xi_2 \right] \xi_2}{\xi_1^2 + \xi_2^2} \right)^{\frac{1}{5}} \right) .$$

Those seemingly mysterious expressions do not make a lot of sense for now. We shall see how they appear in an example in Chap. 7. In fact, they are given here for the sake of completeness but only the expression of Φ_3 and the fact that Φ_4 is continuous matter here. We are interested in estimating trajectories remaining in a given compact set $\mathcal{M}$ which will be defined in Chap. 7.

The function Φ_3 is not Lipschitz at the points on the hyperplanes $\xi_3 = -\xi_1$ and $\xi_1 = 0$. The function Φ_4 is continuous and therefore bounded on any compact set $\mathcal{M}$. Besides, for $\hat{\boldsymbol{\xi}}_3$, $\boldsymbol{\xi}_3$ and u bounded, there exist[11] $c_1 > 0$ and $c_3 > 0$ such that

[11] Let $\Delta\Phi_3(\xi_1, \xi_3, e_1, e_3) = |\xi_3 + e_3 + \xi_1 + e_1|^{\frac{4}{5}} \lfloor \xi_1 + e_1 \rceil^{\frac{1}{5}} - |\xi_3 + \xi_1|^{\frac{4}{5}} \lfloor \xi_1 \rceil^{\frac{1}{5}} = |\xi_3 + \xi_1|^{\frac{4}{5}} \left(\lfloor \xi_1 + e_1 \rceil^{\frac{1}{5}} - \lfloor \xi_1 \rceil^{\frac{1}{5}} \right) + \lfloor \xi_1 + e_1 \rceil^{\frac{1}{5}} \left(|\xi_3 + \xi_1 + e_3 + e_1|^{\frac{4}{5}} - |\xi_3 + \xi_1|^{\frac{4}{5}} \right)$. By Lemma A.5, we have $\left| \lfloor \xi_1 + e_1 \rceil^{\frac{1}{5}} - \lfloor \xi_1 \rceil^{\frac{1}{5}} \right| \leq 2^{\frac{4}{5}} |e_1|^{\frac{1}{5}}$ and $\left| |\xi_3 + \xi_1 + e_3 + e_1|^{\frac{4}{5}} - |\xi_3 + \xi_1|^{\frac{4}{5}} \right| \leq 2^{\frac{1}{5}} (|e_3| + |e_1|)^{\frac{4}{5}} \leq 2^{\frac{1}{5}} (|e_3|^{\frac{4}{5}} + |e_1|^{\frac{4}{5}})$. Besides, $|\xi_1 + e_1|^{\frac{1}{5}} \leq |\xi_1|^{\frac{1}{5}} + |e_1|^{\frac{1}{5}}$, so that for ξ_1 and ξ_3 in compact sets, $|\Delta\Phi_3(\xi_1, \xi_3, e_1, e_3)| \leq c_1|e_1|^{\frac{1}{5}} + c_2|e_3|^{\frac{4}{5}} + c_3|e_1|^{\frac{4}{5}} + c_4|e_1|^{\frac{1}{5}}|e_3|^{\frac{4}{5}} + c_5|e_1|$.

$$|\Phi_3(u, \hat{\xi}_1, \hat{\xi}_2, \hat{\xi}_3) - \Phi_3(u, \xi_1, \xi_2, \xi_3)| \le c_1|\hat{\xi}_1 - \xi_1|^{\frac{1}{5}} + c_3|\hat{\xi}_3 - \xi_3|^{\frac{4}{5}} .$$

This implies that Φ_3 is Hölder with order $\frac{1}{5}$ on $\mathcal{M}$. Hence, the nonlinearities Φ_3 and Φ_4 verify the conditions of Table 4.1. It follows that for L sufficiently large, convergence with an arbitrary small error can be achieved with the high-gain observer (4.4). However, Φ_3 does not verify the conditions of Table 4.2. Thus, there is no theoretical guarantee that the homogeneous observer (4.8) with $d_0 = -1$ will provide exact convergence. On the other hand, the homogeneous cascaded observer (4.10)

$$\begin{cases} \dot{\hat{\xi}}_{11} = \hat{\xi}_{12} - L_1 k_{11} \left\lfloor \hat{\xi}_{11} - y \right\rceil^{\frac{2}{3}} \\ \dot{\hat{\xi}}_{12} = \hat{\xi}_{13} - L_1^2 k_{12} \left\lfloor \hat{\xi}_{11} - y \right\rceil^{\frac{1}{3}} \\ \dot{\hat{\xi}}_{13} \in -L_1^3 k_{13}\, \mathrm{S}(\hat{\xi}_{11} - y) \\ \cdots\cdots\cdots\cdots\cdots\cdots\cdots\cdots\cdots\cdots \\ \dot{\hat{\xi}}_{21} = \hat{\xi}_{22} - L_2 k_{21} \left\lfloor \hat{\xi}_{21} - y \right\rceil^{\frac{3}{4}} \\ \dot{\hat{\xi}}_{22} = \hat{\xi}_{23} - L_2^2 k_{22} \left\lfloor \hat{\xi}_{21} - y \right\rceil^{\frac{1}{2}} \\ \dot{\hat{\xi}}_{23} = \hat{\xi}_{24} + \mathsf{sat}(\mathfrak{g}_3(\hat{\xi}_{11}, \hat{\xi}_{12}, \hat{\xi}_{13}))u - L_2^3 k_{23} \left\lfloor \hat{\xi}_{21} - y \right\rceil^{\frac{1}{4}} \\ \dot{\hat{\xi}}_{24} \in -L_2^4 k_{24}\, \mathrm{S}(\hat{\xi}_{21} - y) \end{cases}$$

converges in finite time according to Theorem 4.4.

In fact, as shown in [6], the differences between those observers appear more strikingly in presence of noise. In particular, the trade-off between final error and noise amplification becomes impossible for the standard high-gain observer. As for the homogeneous observers, the final errors are heavily impacted, though less for the cascaded observer. Indeed, implementing an intermediate homogeneous observer of dimension 3 in the first block enables to obtain much better estimates of the first three states ξ_i, which are then used in the nonlinearity of the second block, thus giving a better estimate of ξ_4. ▲

4.2 General Triangular Form: High-Gain-Kalman Design

A more general triangular form is the following:

Definition 4.3 We call *general continuous triangular form* dynamics of the form

---

By Young's inequality, $|e_1|^{\frac{1}{5}}|e_3|^{\frac{4}{5}} \le \frac{1}{5}|e_1| + \frac{4}{5}|e_3|$, and finally, for e_1 and e_3 in compact sets, $|\Delta\Phi_3(\xi_1, \xi_3, e_1, e_3)| \le \tilde{c}_1|e_1|^{\frac{1}{5}} + \tilde{c}_3|e_3|^{\frac{4}{5}}$.

$$\begin{cases} \dot{\xi}_1 = A_1(u,y)\,\xi_2 + \Phi_1(u,\xi_1) \\ \quad\vdots \\ \dot{\xi}_i = A_i(u,y)\,\xi_{i+1} + \Phi_i(u,\xi_1,\dots,\xi_i) \\ \quad\vdots \\ \dot{\xi}_m = \Phi_m(u,\xi) \end{cases} \quad , \quad y = C_1(u)\xi_1 \tag{4.13}$$

where for all i in $\{1,\dots,m\}$, ξ_i is in $\mathbb{R}^{N_i}$, $\sum_{j=1}^m N_j = d_\xi$, $A_i : \mathbb{R}^{d_u} \times \mathbb{R}^{d_y} \to \mathbb{R}^{N_i \times N_{i+1}}$, $C_1 : \mathbb{R}^{d_u} \to \mathbb{R}^{d_y \times N_1}$, and $\Phi_i : \mathbb{R}^{d_u} \times \mathbb{R}^{\sum_{j=1}^i N_j} \to \mathbb{R}^{N_i}$ are continuous functions. If besides the functions $\Phi_i(u,\cdot)$ are globally Lipschitz on $\mathbb{R}^i$ uniformly in u, then we will say *general Lipschitz triangular form*.

Note that when the values of the functions A_i are constant full-column rank matrices and $C_1(u)$ is the identity function, this form covers the standard triangular form (4.2) if $N_i = N_j$ for all (i, j), and also the forms studied in [9] or [16]. In those cases, a high-gain observer is possible because the system is observable for any input and the functions Φ are triangular and Lipschitz. When the dependence on the input and output is allowed in A_i however, the observability of the system depends on those signals and a high gain is no longer sufficient. In fact, System (4.13) is a combination of both (3.1) and (4.2). It is thus quite natural to combine both Kalman and high-gain designs, as proposed in [7] for the case where $N_i = 1$ for all i, and then in [8] for the general case.

In the following, we denote $y_{\xi_0,u}$ the output at time t of system (4.13) initialized at ξ_0 at time 0, and

$$A(u,y) = \begin{pmatrix} 0 & A_1(u,y) & 0 & \dots & 0 \\ \vdots & & \ddots & \ddots & \vdots \\ & & & & 0 \\ & & & & A_{m-1}(u,y) \\ 0 & & \dots & & 0 \end{pmatrix} \quad , \quad C(u) = (C_1(u)\,,\,0,\dots,\,0)$$

$$\Phi(u,\xi) = \begin{pmatrix} \Phi_1(u,\xi_1) \\ \vdots \\ \Phi_i(u,\xi_1,\dots,\xi_i) \\ \vdots \\ \Phi_m(u,\xi_1,\dots,\xi_m) \end{pmatrix} \quad , \quad \mathscr{L} = \begin{pmatrix} L\,I_{N_1} & 0 & \dots & & 0 \\ 0 & \ddots & & & \\ \vdots & & L^i\,I_{N_i} & & \vdots \\ & & & \ddots & 0 \\ 0 & & \dots & 0 & L^m\,I_{N_m} \end{pmatrix}.$$

Theorem 4.5 ([8]) *Assume the input u is such that*

(a) *there exist positive scalars A_{max} and C_{max} such that for any ξ_0, $t \mapsto A(u(t), y_{\xi_0,u}(t))$ is bounded by A_{max}, and $t \mapsto C(u(t))$ bounded by C_{max},*
(b) *for any ξ_0, the extended input $\nu = (u, y_{\xi_0,u})$ is locally regular for the dynamics*

$$\dot{\chi} = A(u, y_{\xi_0,u})\chi \quad , \quad y = C(u)\chi \tag{4.14}$$

uniformly with respect to ξ_0, i.e., there exist strictly positive real numbers α and L_0 such that for any ξ_0, any $L \geq L_0$ and any $t \geq \frac{1}{L}$,

$$\Gamma_{\nu}^{b}\left(t - \frac{1}{L}, t\right) \geq \alpha L \mathscr{L}^{-2}$$

where Γ_{ν}^{b} is the backward observability grammian[12] *associated to System* (4.14).

(c) *the functions $\Phi_i(u, \cdot)$ are globally Lipschitz on $\mathbb{R}^{\sum_{j=1}^{i} N_j}$ uniformly in u.*

Then, for any positive definite matrix P_0, there exist positive scalars L^*, α_1 and α_2 such that for any $L \geq L^*$, any $\lambda \geq 2A_{max}$ and any $\xi_0 \in \mathbb{R}^{d_\xi}$, the matrix differential equation

$$\dot{P} = L\left(-\lambda P - A(u, y)^{\top} P - P A(u, y) + C(u)^{\top} C(u)\right) \tag{4.15}$$

initialized at $P(0) = P_0$ admits a unique solution verifying $P(t)^{\top} = P(t)$ for all t and

$$\alpha_1 I \leq P(t) \leq \alpha_2 I \qquad \forall t \geq \frac{1}{L}, \tag{4.16}$$

and the system

$$\dot{\hat{\xi}} = A(u, y)\,\hat{\xi} + \Phi(u, \hat{\xi}) + \mathscr{L} P^{-1} C(u)^{\top}\left(y - C(u)\hat{\xi}\right) \tag{4.17}$$

is an observer for the general Lipschitz triangular form (4.13).

Proof The proof consists of a combination of the proofs of Theorems 3.3 and 4.1. See Appendix B.3. □

As opposed to the classical Kalman observer (3.10), the input needs to be more than regularly persistent, namely to be locally regular. This is because in a high gain design, observability at arbitrarily short times is necessary. Note that in the case where the matrices A_i are of dimension one, [15, Lemma 2.1] shows that the gain K can be taken constant under the only condition that there exists A_{min} and A_{max} such that for any ξ_0,

$$0 < A_{min} < A_i(u(t), y_{\xi_0,u}(t)) < A_{max}\ .$$

References

1. Andrieu, V., Praly, L., Astolfi, A.: Nonlinear output feedback design via domination and generalized weighted homogeneity. In: IEEE Conference on Decision and Control (2006)

[12] See Definition 2.1.

2. Andrieu, V., Praly, L., Astolfi, A.: Homogeneous approximation, recursive observer design, and output feedback. SIAM J. Control Optim. **47**(4), 1814–1850 (2008)
3. Andrieu, V., Praly, L., Astolfi, A.: High gain observers with updated gain and homogeneous correction terms. Automatica **45**(2), 422–428 (2009)
4. Astolfi, A., Praly, L.: Global complete observability and output-to-state stability imply the existence of a globally convergent observer. Math. Control Signals Syst. **18**(1), 1–34 (2005)
5. Barbot, J., Boukhobza, T., Djemai, M.: Sliding mode observer for triangular input form. In: IEEE Conference on Decision and Control, vol. 2, pp. 1489–1490 (1996)
6. Bernard, P., Praly, L., Andrieu, V.: Observers for a non-Lipschitz triangular form. Automatica **82**, 301–313 (2017)
7. Besançon, G.: Further results on high gain observers for nonlinear systems. In: IEEE Conference on Decision and Control, vol. 3, pp. 2904–2909 (1999)
8. Besançon, G., Ticlea, A.: An immersion-based observer design for rank-observable nonlinear systems. IEEE Trans. Autom. Control **52**(1), 83–88 (2007)
9. Bornard, G., Hammouri, H.: A high gain observer for a class of uniformly observable systems. In: IEEE Conference on Decision and Control (1991)
10. Cruz-Zavala, E., Moreno, J.A.: Lyapunov functions for continuous and discontinuous differentiators. In: IFAC Symposium on Nonlinear Control Systems (2016)
11. Emelyanov, S., Korovin, S., Nikitin, S., Nikitina, M.: Observers and output differentiators for nonlinear systems. Doklady Akademii Nauk **306**, 556–560 (1989)
12. Filippov, A.: Differential Equations with Discontinuous Right-hand Sides. Mathematics and its applications. Kluwer Academic Publishers Group, Dordrecht (1988)
13. Gauthier, J.P., Bornard, G.: Observability for any u(t) of a class of nonlinear systems. IEEE Trans. Autom. Control **26**, 922–926 (1981)
14. Gauthier, J.P., Hammouri, H., Othman, S.: A simple observer for nonlinear systems application to bioreactors. IEEE Trans. Autom. Control **37**(6), 875–880 (1992)
15. Gauthier, J.P., Kupka, I.: Deterministic Observation Theory and Applications. Cambridge University Press, Cambridge (2001)
16. Hammouri, H., Bornard, G., Busawon, K.: High gain observer for structured multi-output nonlinear systems. IEEE Trans. Autom. Control **55**(4), 987–992 (2010)
17. Khalil, H.K., Praly, L.: High-gain observers in nonlinear feedback control. Int. J. Robust. Nonlinear Control **24** (2013)
18. Levant, A.: Higher-order sliding modes and arbitrary-order exact robust differentiation. In: Proceedings of the European Control Conference, pp. 996–1001 (2001b)
19. Levant, A.: Higher-order sliding modes, differentiation and output-feedback control. Int. J. Control **76**(9–10), 924–941 (2003)
20. Levant, A.: Homogeneity approach to high-order sliding mode design. Automatica **41**(5), 823–830 (2005)
21. Ortiz-Ricardez, F.A., Sanchez, T., Moreno, J.A.: Smooth lyapunov function and gain design for a second order differentiator. In: IEEE Conference on Decision and Control, pp. 5402–5407 (2015)
22. Praly, L., Jiang, Z.: Linear output feedback with dynamic high gain for nonlinear systems. Syst. Control Lett. **53**, 107–116 (2004)
23. Qian, C.: A homogeneous domination approach for global output feedback stabilization of a class of nonlinear systems. In: Proceedings of the American Control Conference (2005)
24. Qian, C.: A homogeneous domination approach for global output feedback stabilization of a class of nonlinear systems. In: IEEE American Control Conference, pp. 4708–4715 (2005)
25. Qian, C., Lin, W.: Recursive observer design, homogeneous approximation, and nonsmooth output feedback stabilization of nonlinear systems. IEEE Trans. Autom. Control **51**(9) (2006)
26. Sanfelice, R., Praly, L.: On the performance of high-gain observers with gain adaptation under measurement noise. Automatica **47**, 2165–2176 (2011)
27. Smirnov, G.: Introduction to the Theory of Differential Inclusions. Graduate studies in mathematics, vol. 41. American Mathematical Society, Providence (2001)

28. Tornambe, A.: Use of asymptotic observers having high gains in the state and parameter estimation. In: IEEE Conference on Decision and Control, vol. 2, pp. 1791–1794 (1989)
29. Yang, B., Lin, W.: Homogeneous observers, iterative design, and global stabilization of high-order nonlinear systems by smooth output feedback. IEEE Trans. Autom. Control **49**(7), 1069–1080 (2004)
30. Zeitz, M.: Observability canonical (phase-variable) form for nonlinear time-variable systems. Int. J. Syst. Sci. **15**(9), 949–958 (1984)

Part II
Transformation into a Normal Form

Chapter 5
Introduction

Throughout Part I, we have given a list of normal forms and their associated observers. We now have to study how a nonlinear system can be transformed into one of those forms to apply Theorem 1.1. This is the object of Part II.

More precisely, we consider a general nonlinear system of the form

$$\dot{x} = f(x, u) \quad , \quad y = h(x, u) \tag{5.1}$$

with x the state in $\mathbb{R}^{d_x}$, u an input function in $\mathscr{U}$ with values in $U \subset \mathbb{R}^{d_u}$, y the output with values in $\mathbb{R}^{d_y}$. For each normal form presented in Part I of the form

$$\dot{\xi} = F(\xi, u, H(\xi, u)) \quad , \quad y = H(\xi, u) \,, \tag{5.2}$$

we look for sufficient conditions on System (5.1) for the existence of a subset $\mathscr{X}$ and functions $T_u : \mathscr{X} \times [0, +\infty[\to \mathbb{R}^{d_\xi}$ for each u in $\mathscr{U}$ which transforms System (5.1) into the normal form (5.2) in the sense of Theorem 1.1, i.e., for all x in $\mathscr{X}$ and all t in $[0, +\infty)$,[1]

$$L_{(f,1)}T_u(x, t) = F(T_u(x, t), u(t), h(x, u(t))) \quad , \quad h(x, u(t)) = H(T_u(x, t), u(t)) \,.$$

Indeed, according to Theorem 1.1 and Corollary 1.1, the observer design problem is then solved for System (5.1) if the solutions of System (5.1) which are of interest remain in $\mathscr{X}$ and

- *either* for any u in $\mathscr{U}$, $x \mapsto T_u(x, t)$ becomes injective on $\mathscr{X}$ uniformly in space and in time after a certain time;
- *or* $\mathscr{C} = \mathscr{X}$ is a compact set, and for any u in $\mathscr{U}$, T_u is a same stationary transformation T injective on $\mathscr{C}$.

[1]With $L_{(f,1)}T_u(x, t) = \lim_{h\to 0} \frac{T_u(X(x,t;t+h;u),t+h)-T_u(x,t)}{h}$.

P. Bernard, *Observer Design for Nonlinear Systems*,
Lecture Notes in Control and Information Sciences 479,
https://doi.org/10.1007/978-3-030-11146-5_5

In order to be transformable into one of the normal forms, System (5.1) will need to verify some observability assumptions. Some of them have already been defined in Definition 1.2, but another crucial notion (quickly mentioned in the introduction) is the so-called differential observability, which roughly says that the map made of the input and some of its derivatives contains all the information about the state, namely is injective. In order to properly define this map, we need the following definition.

Definition 5.1 Given an integer m, and using the notation

$$\overline{\nu}_m = (\nu_0, \dots, \nu_m) ,$$

we call *dynamic extension of order m* of System (5.1) the extended dynamical system

$$\dot{\overline{x}} = \overline{f}(\overline{x}, u^{(m+1)}) \quad , \quad y = \hbar(\overline{x}) \tag{5.3}$$

with input $u^{(m+1)}$ in $\mathbb{R}^{d_u}$, extended state $\overline{x} = (x, \overline{\nu}_m)$ in $\mathbb{R}^{d_x} \times \mathbb{R}^{d_u(m+1)}$, extended vector field $\overline{f}$ defined by

$$\overline{f}(\overline{x}, u^{(m+1)}) = \left(f(x, \nu_0)\, , \, \nu_1\, , \, \dots\, , \, \nu_m\, , \, u^{(m+1)}\right)$$

and extended measurement function $\hbar$ defined by

$$\hbar(\overline{x}) = h(x, \nu_0) .$$

Note that for any solution x to System (5.1) with some input u, $(x, \overline{u}_m)$ is solution to the dynamic extension (5.3), with the notation $\overline{u}_m = (u, \dot{u}, \dots, u^{(m)})$. While $\overline{\nu}_m$ is an element of $\mathbb{R}^{d_u(m+1)}$, $\overline{u}_m$ is a function defined on $[0, +\infty)$ such that $\overline{u}_m(s) = (u(t), \dot{u}(t), \dots, u^{(m)}(t))$ is in $\overline{U}_m \subset \mathbb{R}^{d_u(m+1)}$. The successive time derivatives of the output y are related to the Lie derivatives of $\hbar$ along the vector fields $\overline{f}$, namely for any $j \leq m$ and any (x_0, t_0) in $\mathcal{X} \times [0, +\infty)$

$$\frac{\partial^j Y}{\partial t^j}(x_0, t_0; t; u) = L_{\overline{f}}^j \hbar(X(x_0, t_0; t; u), \overline{u}_m(t)) .$$

We are now ready to define the notion of differential observability.

Definition 5.2 Consider the function $\overline{\mathbf{H}}_m$ on $\mathbb{R}^{d_x} \times \mathbb{R}^{d_u(m+1)}$ defined by

$$\overline{\mathbf{H}}_m(x, \overline{\nu}_m) = \left(\hbar(x, \overline{\nu}_m)\, , \, L_{\overline{f}}\hbar(x, \overline{\nu}_m)\, , \, \dots\, , \, L_{\overline{f}}^{m-1}\hbar(x, \overline{\nu}_m)\right) . \tag{5.4}$$

System (5.1) is

- *weakly differentially observable of order m on $\mathcal{S}$* if for any $\overline{\nu}_m$ in $\overline{U}_m$, the function $x \mapsto \overline{\mathbf{H}}_m(x, \overline{\nu}_m)$ is injective on $\mathcal{S}$.
- *strongly differentially observable of order m on $\mathcal{S}$* if for any $\overline{\nu}_m$ in $\overline{U}_m$, $x \mapsto \overline{\mathbf{H}}_m(x, \overline{\nu}_m)$ is an injective immersion on $\mathcal{S}$.

The notion of differential observability of order m thus means that when knowing the current input and its derivatives, the current state is uniquely determined by the current output and its first $m-1$ derivatives.

In some cases, it will be useful to restrict our attention to control-affine multi-input single-output systems of the form

$$\dot{x} = f(x) + g(x)u \quad , \quad y = h(x) \in \mathbb{R} \tag{5.5}$$

with $g : \mathbb{R}^{d_x} \mapsto \mathbb{R}^{d_x \times d_u}$, for which we introduce the following definition.

Definition 5.3 We call *drift system* of System (5.5) the dynamics with $u \equiv 0$, namely

$$\dot{x} = f(x) \quad , \quad y = h(x)\,.$$

Applying Definition 5.2, we say that the drift system of System (5.5) is weakly (resp strongly) differentially observable of order m on $\mathscr{S}$ if the function

$$\mathbf{H}_m(x) = (h(x), L_f h(x), \ldots, L_f^{m-1} h(x)) \tag{5.6}$$

is injective (resp an injective immersion) on $\mathscr{S}$.

Differential observability of the drift system is weaker than differential observability of the system since it is only for $u \equiv 0$.[2]

With those definitions in hand, we are now ready to introduce the main results available in the literature concerning the transformation of nonlinear systems into state-affine normal forms in Chap. 6, and then into triangular normal forms in Chap. 7. Those results are summed up in Table 5.1.

[2]Or any other constant value.

Table 5.1 Which type of nonlinear system, under which observability condition and with which transformation and domain of validity can be transformed into each of the normal forms presented in Part I

Normal form		Type of system	Observability assumption	Transformation T	T time-varying?	Validity
State-affine form (3.4) or (3.6)		/	Diverse but very restrictive	Variable	No	Often only local
Hurwitz form (3.2)		Autonomous	Backward distinguishability	Solution to PDE (6.5)	No	Global
		Controlled	Backward distinguishability	Solution to PDE (6.7)	Yes, dependent on the input trajectory u	Global, injectivity after a certain time
Triangular form (4.2)	Continuous phase-variable	/	Weakly observable of order m	Output and its derivatives up to order $m-1$	Yes, with derivatives of input	At least on any compact set
	Lipschitz phase-variable	/	Strongly observable of order m	Output and its derivatives up to order $m-1$	Yes, with derivatives of input	At least on any compact set
	Continuous triangular $d_y = 1$	Control-affine single-output	Conjecture: uniformly observable and weakly differentially observable of order $d_\xi \geq d_x$	Output and its derivatives along drift vector field up to order $d_\xi - 1$	No	At least on compact sets
	Lipschitz triangular $d_y = 1$	Control-affine single-output	Uniformly observable and strongly differentially observable of order d_x	Output and its derivatives along drift vector field up to order $d_x - 1$	No	At least on compact sets
General Lipschitz triangular form (4.14)		Control-affine single-output	Observability rank condition	Output and its derivatives along each vector field	No	At least locally

Chapter 6
Transformations into State-Affine Normal Forms

In this chapter, we look for sufficient conditions on a nonlinear system ensuring the existence of a transformation into one of the state-affine normal forms presented in Chap. 3. To avoid unpleasant ruptures in the reading, the lengthy proofs are given in Appendix C.

6.1 Linearization by Output Injection

6.1.1 Constant Linear Part

The problem of transforming a nonlinear system into a linear one of the form (3.4), i.e.,

$$\dot{\xi} = A\,\xi + B(u, \tilde{y}), \quad \tilde{y} = \psi(y) = C\,\xi \tag{6.1}$$

with the pair (A, C) observable and $\psi : \mathbb{R}^{d_y} \to \mathbb{R}^{d_y}$ a possible change of output, has a very long history. The first results appeared in [7, 23] for autonomous systems and were then extended by [26] to multi-input multi-output systems. In those papers, the authors looked for necessary and sufficient conditions on the functions f and h for the existence of a local change of coordinates (and possibly change of output) which brings the system into the form (6.1), which they called "observer form" [8], and then gave conditions for the existence of a local (and global) immersion[1] (instead of diffeomorphism) in the particular case of control-affine systems. A vast literature followed on the subject, either developing algebraic algorithms to check the existence of a transformation or tools to explicitly find the transformation.

[1] $T : \mathbb{R}^{d_x} \to \mathbb{R}^{d_\xi}$ is an immersion if the rank of $\frac{\partial T}{\partial x}$ is d_x. Contrary to a diffeomorphism, this allows to take $d_\xi \geq d_x$.

P. Bernard, *Observer Design for Nonlinear Systems*,
Lecture Notes in Control and Information Sciences 479,
https://doi.org/10.1007/978-3-030-11146-5_6

In [19], the general question of existence/construction of a transformation (not necessarily injective/immersion/diffeomorphism[2]) into the form (6.1) (without even the assumption of observability of the pair (A, C)) is addressed. If such a transformation exists, the system is said linearizable by output injection. The following result is proved.

Theorem 6.1 ([19]) *A system of the form*

$$\dot{x} = f(x, u), \quad y = h(x)$$

is linearizable by output injection if and only if there exist a map T and a diffeomorphism $\psi : \mathbb{R}^{d_y} \to \mathbb{R}^{d_y}$ transforming the system into the canonical triangular form

$$\begin{cases} \dot{\xi}_1 &= \xi_2 + \Phi_1(u, \xi_1) \\ &\vdots \\ \dot{\xi}_i &= \xi_{i+1} + \Phi_i(u, \xi_1) \\ &\vdots \\ \dot{\xi}_{d_\xi} &= \Phi_{d_\xi}(u, \xi_1) \end{cases} , \quad \tilde{y} = \psi(y) = \xi_1 ,$$

if $d_y = 1$, and d_y blocks of such triangular forms if $d_y > 1$.

Proof. Assume the system is linearizable by output injection (the converse is immediate) with maps T_0 and ψ_0, and pair (A_0, C_0). Based on well-known facts on the reduction of linear systems, by considering the composition of T_0 with the projection on the observable space of (A_0, C_0) (nonempty as long as $C_0 \neq 0$), and keeping ψ_0, we get a linear structure (of smaller dimension) with an observable pair (A, C). Thanks to a linear change of coordinates, we know that (A, C) can be transformed into d_y blocks of

$$A_c = \begin{pmatrix} * & 1 & 0 & \dots & 0 \\ * & 0 & 1 & & 0 \\ \vdots & \vdots & \ddots & \ddots & \\ * & & & & 1 \\ * & 0 & & \dots & 0 \end{pmatrix}, \quad C_c = \begin{pmatrix} 1 & 0 & \dots & 0 \end{pmatrix}$$

The first column of A_c represents terms that depend on the outputs only and can be incorporated into Φ, therefore giving the result. □

Thus, the linearization problem reduces to the existence of a transformation into this latter observable form. Note that if besides this transformation is required to be injective (like in our context of observer design), then the system is necessarily uniformly observable, i.e., observable for any input.[3] Actually, the class of systems

[2] In [19], the transformation is called an *immersion*, but as defined in [19, Definition 2.2], and not in the usual differential geometry sense as defined in this book (see previous footnote).

[3] See Definition 1.2.

considered here is even strictly smaller because for a uniformly observable system, the functions Φ_i would be allowed to depend on $\xi_1, \ldots, \xi_i$, and not only on ξ_1.

From this, it is possible to deduce:

Theorem 6.2 ([19]) *An autonomous system*

$$\dot{x} = f(x), \quad y = h(x) \in \mathbb{R}$$

is linearizable by output injection if and only if there exist a function ψ, an integer d_ξ, and functions Φ_1, ..., Φ_{d_ξ} such that

$$L_f^{d_\xi} \tilde{h} = L_f^{d_\xi - 1} \Phi_1 \circ \tilde{h} + L_f^{d_\xi - 2} \Phi_2 \circ \tilde{h} + \ldots + L_f \Phi_{d_\xi - 1} \circ \tilde{h} + \Phi_{d_\xi} \circ \tilde{h} \tag{6.2}$$

with $\tilde{h} = \psi \circ h$.

Proof. Assume the system is linearizable by output injection with map $T = (T_1, \ldots, T_{d_\xi})$ and output diffeomorphism ψ transforming the system into the canonical triangular form given by Theorem 6.1. Take $\tilde{h} = \psi \circ h = T_1$. By induction, for $1 \leq k < d_\xi$, assume

$$T_k = L_f^k \tilde{h} - L_f^{k-1} \Phi_1 \circ \tilde{h} - L_f^{k-2} \Phi_2 \circ \tilde{h} - \ldots - L_f \Phi_{k-2} \circ \tilde{h} - \Phi_{k-1} \circ \tilde{h} \, ,$$

then

$$\begin{aligned} T_{k+1} &= L_f T_k - \Phi_k \circ \tilde{h} \\ &= L_f^{k+1} \tilde{h} - L_f^k \Phi_1 \circ \tilde{h} - L_f^{k-1} \Phi_2 \circ \tilde{h} - \ldots - L_f \Phi_{k-1} \circ \tilde{h} - \Phi_k \circ \tilde{h} \end{aligned}$$

and

$$\begin{aligned} \Phi_{d_\xi} \circ \tilde{h} &= L_f T_{d_\xi} \\ &= L_f^{d_\xi} \tilde{h} - L_f^{d_\xi - 1} \Phi_1 \circ \tilde{h} - L_f^{d_\xi - 2} \Phi_2 \circ \tilde{h} - \ldots - L_f \Phi_{d_\xi - 1} \circ \tilde{h} \end{aligned}$$

which gives (6.2). Conversely, assume that (6.2) holds. Then, T defined by

$$\begin{cases} T_1 = \tilde{h} \\ T_{k+1} = L_f T_k - \Phi_k \circ \tilde{h} \; 1 \leq k < d_\xi \end{cases}$$

transforms the system into the canonical form of Theorem 6.1. □

Equation (6.2) is the so-called characteristic equation which extends the same notion for linear systems and was introduced in [21] originally with $d_\xi = d_x$. When $d_y > 1$, we get d_y characteristic equations. This partial differential equation (PDE) is important in practice because several results show that the linearization of a controlled system first necessitates the linearization of its uncontrolled parts or drift dynamics[4] ([8, 19, 26] among others). A first difficulty thus lies in solving this PDE, which does not always admit solutions ([3, 19]).

Along the history of linearization, we must also mention some generalizations such as [21], where the function B is allowed to depend on the derivatives of the input and later on the derivatives of the output in [13, 28], or [14, 29] where it is proposed to use an output-depending timescale transformation.

We conclude that linearizing both the dynamics and the output function is very demanding and requires some very restrictive conditions on the system. The existence of the transformation is difficult to check and involves quite tedious symbolic calculations which do not always provide the transformation itself, and even when they do, its validity is often only local.

6.1.2 Time-Varying Linear Part

In parallel, others allowed the linear part A to depend on the input/output, i.e., looked for conditions to transform the system into the state-affine form (3.6)

$$\dot{\xi} = A(u, y)\,\xi + B(u, y), \quad y = C(u)\xi \;.$$

The first to address this problem were [11, 12] but without allowing output injection in the dynamics, namely requiring $A(u)$ and $B(u)$. This led to the very restrictive finiteness criterion of the observation space, which roughly says that the linear space containing the successive derivatives of the output along any vector field of the type $f(\cdot, u)$ is finite. Later, [6, 16, 17] allowed A and B to depend on the output to broaden the class of concerned systems. But those systems remain difficult to characterize because there are often many possible ways to parametrize the system via the output. Besides, even when the transformation exists and is known, the input must satisfy an extra excitation condition to allow the design of a Kalman observer (see Chap. 3).

6.2 Transformation into Hurwitz Form

When Luenberger published his first results concerning the design of observers for linear systems in [27], his idea consisted in transforming the linear plant

$$\dot{x} = F\,x, \quad y = C\,x$$

into a Hurwitz form

$$\dot{\xi} = A\,\xi + B\,y \;. \tag{6.3}$$

[4]Dynamics with u equal to a constant.

Indeed, as we saw in Chap. 3, this system admits a trivial observer made of a copy of the dynamics. He proved that when the pair (F, C) is observable, this is always possible via a linear stationary transformation $\xi = T\,x$ with $d_\xi = d_x$, with A any Hurwitz matrix in $\mathbb{R}^{d_x \times d_x}$ and B any vector in $\mathbb{R}^{d_x \times d_y}$ such that the pair (A, B) is controllable. This is based[5] on the fact that the Sylvester equation

$$TF = AT + BC \tag{6.4}$$

admits in this case a solution that is unique and invertible.

Some researchers have therefore tried to reproduce Luenberger's original methodology on nonlinear systems, i.e., find a transformation into a Hurwitz form (3.2)

$$\dot{\xi} = A\,\xi + B(u, y), \quad y = H(\xi, u)$$

with A Hurwitz. Unlike in the previous section, this procedure is not a linearization of the system, since the output function H can be any nonlinear function (see [20, Remark 4]). This crucial difference leads to far less restrictive conditions on the system, and because the corresponding observer (3.3) is a simple copy of the dynamics, it is not even necessary to have an explicit expression for H.

6.2.1 Luenberger Design for Autonomous Systems

The extension of this Luenberger design from linear systems to autonomous nonlinear systems was first proposed and analyzed in a general context by [31]. It was rediscovered later by [20] who gave a local analysis close to an equilibrium point under conditions relaxed later on in [24]. The localness and most of the restrictive assumptions were then bypassed in [2]. As noticed in [2, 25], this nonlinear Luenberger observer is also strongly related to the observer proposed in [22].

In [2], the authors investigate the possibility of transforming an autonomous system

$$\dot{x} = f(x), \quad y = h(x)$$

into a Hurwitz form

$$\dot{\xi} = A\,\xi + B(y)\,.$$

This raises the question of finding, for some integer d_ξ, a continuous function $T : \mathbb{R}^{d_x} \to \mathbb{R}^{d_\xi}$ verifying

$$L_f T(x) = A\,T(x) + B(h(x)), \quad \forall x \in \mathcal{X} \tag{6.5}$$

[5] $\xi = T\,x$ follows the dynamics (6.3) if and only if (6.4).

with A some Hurwitz matrix of dimension d_ξ and $B : \mathbb{R}^{d_y} \to \mathbb{R}^{d_\xi}$ some continuous function. The existence of such a transformation is shown for any Hurwitz matrix A and for some well-chosen functions B, under the only assumption that the system is backward complete[6] in $\mathscr{X}$ ([2, Theorem 2]). Of course, this is not enough since, as we saw in the introduction, it is required that T be uniformly injective on $\mathscr{X}$ to deduce from the estimate of $\xi = T(x)$ an estimate of x. The authors show in [2, Theorem 3] that injectivity of T is achieved for almost any diagonal complex Hurwitz matrix A of dimension[7] $(d_x + 1)d_y$ on $\mathbb{C}$ and for any B verifying some growth condition under the assumption that the system is backward $\mathscr{S}$-distinguishable[8] on $\mathscr{X}$ for some open set $\mathscr{S}$ containing $\mathscr{X}$; i.e., for any (x_a, x_b) in $\mathscr{X}^2$ such that $x_a \neq x_b$, there exists t in $\left(\max\left\{\sigma_{\mathscr{S}}^-(x_a), \sigma_{\mathscr{S}}^-(x_b)\right\}, 0\right]$ such that $y_{x_a}(t) \neq y_{x_b}(t)$.

In the case where $\mathscr{X}$ is bounded, the result simplifies into:

Theorem 6.3 ([2]) *Assume that $\mathscr{X}$ and $\mathscr{S}$ are open bounded subsets of $\mathbb{R}^{d_x}$, such that $\mathtt{cl}(\mathscr{X})$ is contained in $\mathscr{S}$ and System (5.1) is backward $\mathscr{S}$-distinguishable on $\mathscr{X}$. There exists a strictly positive number ℓ and a set $\mathscr{R}$ of zero Lebesgue measure in $\mathbb{C}^{d_x+1}$ such that denoting $\Omega = \{\lambda \in \mathbb{C} : \Re(\lambda) < -\ell\}$, for any $(\lambda_1, \dots, \lambda_{d_x+1})$ in $\Omega^{d_x+1} \setminus \mathscr{R}$, there exists a function $T : \mathbb{R}^{d_x} \to \mathbb{R}^{(d_x+1)\times d_y}$ uniformly injective on $\mathscr{X}$ and verifying* (6.5) *with*

$$A = \tilde{A} \otimes I_{d_y}, \quad B(y) = \tilde{B} \otimes I_{d_y}\, y$$

and

$$\tilde{A} = \begin{pmatrix} \lambda_1 & & \\ & \ddots & \\ & & \lambda_{d_x+1} \end{pmatrix}, \quad \tilde{B} = \begin{pmatrix} 1 \\ \vdots \\ 1 \end{pmatrix}.$$

Besides, if $\mathscr{X}$ is backward invariant, the function T is unique on $\mathscr{X}$ and defined by

$$T(x) = \int_{-\infty}^{0} e^{-A\tau} B(h(X(x,\tau)))\, d\tau\,. \tag{6.6}$$

Remark 6.1 The function T proposed by Theorem 6.3 takes complex values. To remain in the real frame, one should consider the transformation made of its real and imaginary parts, and instead of implementing for each i in $\{1, \dots, d_y\}$ and each λ

$$\dot{\hat{\xi}}_{\lambda,i} = -\lambda\hat{\xi}_\lambda + y_i$$

[6] Any solution exiting $\mathscr{X}$ in finite time must cross the boundary of $\mathscr{X}$. See [2, Definition 1].

[7] Separating the real/imaginary parts, the observer is thus of dimension $2(d_x + 1)d_y$ on $\mathbb{R}$. See Remark 6.1.

[8] This notion is similar to the distinguishability defined in Definition 1.2 but in negative time and with the constraint that $\bar{t}$ occurs when both solutions are still in $\mathscr{S}$.

in $\mathbb{C}$, one should implement

$$\dot{\hat{\xi}}_{\lambda,i} = \begin{pmatrix} -\Re(\lambda) & -\Im(\lambda) \\ \Im(\lambda) & -\Re(\lambda) \end{pmatrix} \hat{\xi}_{\lambda,i} + \begin{pmatrix} y_i \\ 0 \end{pmatrix}$$

in $\mathbb{R}$. Thus, the dimension of the observer is $2 \times d_y \times (d_x + 1)$ in terms of real variables.

We conclude from this result that it is possible to design an observer for an autonomous nonlinear system under the weak assumption of backward distinguishability. Note that with a stronger assumption of strong differential observability[9] of order m, and still in a bounded set, it is also proved in [2, Theorem 4] that injectivity of (6.6) is ensured for any choice of m real strictly negative λ_i smaller that $-\ell$ with ℓ sufficiently large.

The difficulty lies in the computation of the function T, let alone its inverse. Even when $\mathscr{X}$ is bounded and backward invariant, the use of its explicit expression (6.6) is not easy since it necessitates to integrate backward the differential equation at each time step. Several examples will be given in Sect. 6.3 or Chap. 11 to show how the function T can be computed without relying on this formula. In particular, we will see that even for an autonomous system, this task can sometimes be made easier by allowing T to be time-varying.

6.2.2 Luenberger Design for Nonautonomous Systems

After first steps in [30, 32] for linear time-varying systems and in [15] for nonlinear time-varying systems, the extension of the Luenberger design to general controlled systems was considered in [9, 10], following the ideas of [22], and more recently[10] in [4, 5].

In fact, when we want to extend the previous results to time-varying/controlled systems, two paths are possible: either we keep the stationary transformation obtained for some constant value of u (for instance the drift system at $u \equiv 0$) and hope that the additional terms due to the presence of u do not prevent convergence, or we take a time-varying transformation taking into account (implicitly or explicitly) the time or input u. The idea pursued in [9] belongs to the first path: The transformation is stationary, and the input is seen as a disturbance which must be small enough. Although the construction is extended in a cunning fashion to a larger class of inputs, namely those which can be considered as output of a linear generator model with small external input, this approach remains theoretical and restrictive. On the other hand, in [10], the author rather tries to use a time-varying transformation but its injectivity is proved only under the so-called finite complexity assumption, originally

[9] See Definition 5.2 in the autonomous case.

[10] Some texts of Sect. 6.2.2 are reproduced from [5] with permission from IEEE.

introduced in [22] for autonomous systems. Unfortunately, this property is very restrictive and hard to check. Besides, no indication about the dimension d_ξ is given and the transformation cannot be computed online because it depends on the whole past trajectory of the output. Those problems were then overcome in [4, 5] with results of existence and injectivity of a time-varying transformation under more standard and constructive observability assumptions.

In order to transform System (5.1) into a Hurwitz form[11] (6.3) with A Hurwitz in $\mathbb{R}^{d_\xi \times d_\xi}$, B a vector in $\mathbb{R}^{d_\xi \times d_y}$, for some strictly positive integer d_ξ, we need to find (for each u in $\mathscr{U}$) a transformation[12] $T : \mathbb{R}^{d_x} \times [0, +\infty) \to \mathbb{R}^{d_\xi}$ such that for any x in $\mathscr{X}$ and any time t in $[0, +\infty)$,

$$\frac{\partial T}{\partial x}(x, t) f(x, u(t)) + \frac{\partial T}{\partial t}(x, t) = A\,T(x, t) + B\,h(x, u(t)) \,. \tag{6.7}$$

According to Theorem 1.1, if besides T becomes injective uniformly in time and in space at least after a certain time, this will give an observer for System (5.1).

6.2.2.1 Existence of a Time-Varying Transformation

The existence of a C^1 time-varying solution to PDE (6.7) is achieved under the following assumption.

Assumption 6.1 *System (5.1) does not blow up in finite backward time; namely for any u in $\mathscr{U}$, any (x, t) in $\mathscr{X} \times [0, +\infty)$, and any s in $[0, t]$, $X(x, t; s; u)$ is defined.*

The result reads as follows.

Lemma 6.1 ([4, 5]) *Consider a strictly positive number d_ξ, a Hurwitz matrix A in $\mathbb{R}^{d_\xi \times d_\xi}$, a matrix B in $\mathbb{R}^{d_\xi \times d_y}$, and an input u in $\mathscr{U}$. Under Assumption 6.1, the function T^0 defined on $\mathscr{X} \times [0, +\infty)$ by*

$$T^0(x, t) = \int_0^t e^{A(t-s)} B\,Y(x, t; s; u)\,ds \tag{6.8}$$

is a C^1 solution to PDE (6.7) on $\mathscr{X} \times [0, +\infty)$.

Proof. First, for any u in $\mathscr{U}$, and any s in $[0, +\infty)$, $(x, t) \mapsto Y(x, t; s; u) = h(X(x, t; s; u), s)$ is C^1; thus T^0 is C^1. Take x in $\mathscr{X}$ and t in $[0, +\infty)$. For any τ in $\mathbb{R}$,

$$X(X(x, t, t + \tau; u), t + \tau; s; u) = X(x, t; s; u).$$

[11] We could have considered a more general Hurwitz form $\dot{\xi} = A\,\xi + B(y)$ with B any nonlinear function, but taking B linear is sufficient to obtain satisfactory results.

[12] The function T depends on u in $\mathscr{U}$, and we should write T_u as in Theorem 1.1. But we drop this too heavy notation in this chapter to ease the comprehension. What is important is that the target Hurwitz form, namely d_ξ, A, and B, is the same for all u in $\mathscr{U}$.

Therefore,

$$\begin{aligned} T^0(X(x,t;t+\tau;u),t+\tau) &= \int_0^{t+\tau} e^{A(t+\tau-s)} Bh(X(x,t,s;u),u(s))ds \\ &= e^{A\tau}\, T^0(x,t) + e^{A\tau} \int_t^{t+\tau} e^{A(t-s)} B\, h(X(x,t;s;u),u(s))ds \end{aligned}$$

and

$$\begin{aligned} \frac{T^0(X(x,t;t+\tau;u),t+\tau) - T^0(x,t)}{\tau} &= \frac{e^{A\tau}-I}{\tau} T^0(x,t) \\ &\quad + \frac{e^{A\tau}}{\tau} \int_t^{t+\tau} e^{A(t-s)} B\, h(X(x,t,s;u),u(s))ds\ . \end{aligned}$$

Making τ tend to 0, we get PDE (6.7). □

Note that it may still be possible to construct a function T^0 solution to PDE (6.7) on $\mathscr{X} \times [0,+\infty)$ when Assumption 6.1 does not hold. This is the case if there exists a subset $\mathscr{X}'$ of $\mathbb{R}^{d_x}$ such that $\mathtt{cl}(\mathscr{X}) \subset \mathscr{X}'$ and which any blowing-up solution in backward time has to leave.[13] Indeed, any modified dynamics

$$\dot{x} = \chi(x) f(x,u)\ , \tag{6.9}$$

with a C^∞ function $\chi : \mathbb{R}^{d_x} \to \mathbb{R}$ satisfying

$$\chi(x) = \begin{cases} 1\ , & \text{if } x \in \mathtt{cl}(\mathscr{X}) \\ 0\ , & \text{if } x \notin \mathscr{X}' \end{cases}$$

is then backward complete and satisfies Assumption 6.1. Besides, the PDE associated with system (6.9) is the same as PDE (6.7) on $\mathscr{X} \times [0,+\infty)$. This allows us to deduce the following corollary.

Corollary 6.1 ([5]) *Assume $\mathscr{X}$ is bounded. Consider a strictly positive number d_ξ, a Hurwitz matrix A in $\mathbb{R}^{d_\xi \times d_\xi}$, a matrix B in $\mathbb{R}^{d_\xi \times d_y}$, and an input u in $\mathscr{U}$. There exists a C^1 function T^0 solution to PDE (6.7) on $\mathscr{X} \times [0,+\infty)$.*

Observe that T^0 depends only on the values of the input u on $[0,t]$, so that it is theoretically computable online. However, for each couple (x,t), one would need to integrate backward the dynamics (5.1) until time 0, which is quite heavy. If the input u is known in advance (for instance $u(t)=t$), it can also be computed on a grid offline. We will see in Sect. 6.3 on practical examples how we can find a solution to PDE (6.7) without relying on the expression T^0.

In any case, we conclude that a C^1 time-varying transformation into a Hurwitz form always exists under the mild Assumption 6.1, but the core of the problem is to ensure its injectivity.

---

[13] This property is named completeness within $\mathscr{X}'$ in [2].

6.2.2.2 Injectivity with Strong Differential Observability

We will need the following assumption which uses the dynamic extension defined in Definition 5.1.

Assumption 6.2 *There exists a subset $\mathcal{S}$ of $\mathbb{R}^{d_x}$ such that:*

1. *For any u in $\mathcal{U}$, any x in $\mathcal{S}$, and any time t in $[0, +\infty)$, $X(x, t; s; u)$ is in $\mathcal{S}$ for all s in $[0, +\infty)$.*
2. *The quantity $M_f := \sup_{\substack{x \in \mathcal{S} \\ \nu_0 \in U}} \left| \frac{\partial f}{\partial x}(x, \nu_0) \right|$ is finite.*
3. *There exist d_y integers $(m_1, \dots, m_{d_y})$ such that the functions*

$$H_i(x, \overline{\nu}_m) = \left(\bar{h}_i(x, \overline{\nu}_m), L_{\bar{f}}\bar{h}_i(x, \overline{\nu}_m), \ \dots \ , L_{\bar{f}}^{m_i-1}\bar{h}_i(x, \overline{\nu}_m)\right) \tag{6.10}$$

   defined on $\mathcal{S} \times \mathbb{R}^{d_u(m+1)}$ with $m = \max_i m_i$ and $1 \leq i \leq d_y$ verify:

   – *For all u in $\mathcal{U}$, $H_i(\cdot, \overline{u}_m(0)))$ is Lipschitz on $\mathcal{S}$.*
   – *There exists L_H such that the function*

$$H(x, \overline{\nu}_m) = \left(H_1(x, \overline{\nu}_m), \ \dots \ , H_i(x, \overline{\nu}_m), \ \dots \ , H_{d_y}(x, \overline{\nu}_m)\right) \tag{6.11}$$

   verifies for any (x_1, x_2) in $\mathcal{S}^2$ and any $\overline{\nu}_m$ in $\overline{U}_m$

$$|x_1 - x_2| \leq L_H |H(x_1, \overline{\nu}_m) - H(x_2, \overline{\nu}_m)|$$

   Namely, H is Lipschitz injective on $\mathcal{S}$, uniformly with respect to $\overline{\nu}_m$ in $\overline{U}_m$.

4. *For all $1 \leq i \leq d_y$, there exists L_i such that for all (x_1, x_2) in $\mathcal{S}^2$ and for all $\overline{\nu}_m$ in $\overline{U}_m$,*

$$|L_{\bar{f}}^{m_i}\bar{h}_i(x_1, \overline{\nu}_m) - L_{\bar{f}}^{m_i}\bar{h}_i(x_2, \overline{\nu}_m)| \leq L_i|x_1 - x_2|$$

   Namely, $L_{\bar{f}}^{m_i}\bar{h}_i(\cdot, \overline{\nu}_m)$ is Lipschitz on $\mathcal{S}$, uniformly with respect to $\overline{\nu}_m$ in $\overline{U}_m$.

Those assumptions can be simplified in the case where $\mathcal{S}$ is compact.

Lemma 6.2 ([4]) *Assume that $\mathcal{S}$ is compact and there exist d_y integers $(m_1, \dots, m_{d_y})$ such that $\overline{U}_m$ with $m = \max_i m_i$ is compact, and for any $\overline{\nu}_m$ in $\overline{U}_m$, $H(\cdot, \overline{\nu}_m)$ defined in (6.11) is an injective immersion[14] on $\mathcal{S}$. Then, the points 2–3–4 in Assumption 6.2 are satisfied.*

Proof. Consequence of Lemma A.12. □

[14] $H(\cdot, \overline{\nu}_m)$ is injective on $\mathcal{S}$, and $\frac{\partial H}{\partial x}(x, \overline{\nu}_m)$ is full-rank for any x in $\mathcal{S}$.

Note that $H(\cdot, \overline{v}_m)$ defined in (6.11) is similar to $\overline{\mathbf{H}}_m$ defined in Definition 5.2. The only difference is that the orders of differentiation of each component of h are different: If $m_i = m$ for all i in $\{1, \dots, d_y\}$, we recover $\overline{\mathbf{H}}_m$. Therefore, the fact that $H(\cdot, \overline{v}_m)$ is injective (resp. an injective immersion) is a kind of weak (resp. strong) differential observability property as defined in Definition 5.2.

Under Assumption 6.2, the following result was presented in [4].

Theorem 6.4 ([4]) *Suppose Assumption 6.2 holds. Consider Hurwitz matrices A_i in $\mathbb{R}^{m_i \times m_i}$, with m_i defined in Assumption 6.2, and vectors B_i in $\mathbb{R}^{m_i}$ such that the pairs (A_i, B_i) are controllable. There exists a strictly positive real number $\overline{k}$ such that for all $k \geq \overline{k}$, for all input u in $\mathscr{U}$, there exists $\overline{t}_{k,u}$ such that any C^1 solution T to PDE* (6.7) *on $\mathscr{S} \times [0, +\infty)$ with*

- $d_\xi = \sum_{i=1}^{d_y} m_i$
- *A in $\mathbb{R}^{d_\xi \times d_\xi}$ and B in $\mathbb{R}^{d_\xi \times d_y}$ defined by*

$$A = \begin{pmatrix} kA_1 & & & & \\ & \ddots & & & \\ & & kA_i & & \\ & & & \ddots & \\ & & & & kA_{d_y} \end{pmatrix} \qquad B = \begin{pmatrix} B_1 & & & & \\ & \ddots & & & \\ & & B_i & & \\ & & & \ddots & \\ & & & & B_{d_y} \end{pmatrix}$$

- *$T(\cdot, 0)$ Lipschitz on $\mathscr{S}$*

is such that $T(\cdot, t)$ is injective on $\mathscr{S}$ for all $t \geq \overline{t}_{k,u}$, uniformly in time and in space. More precisely, there exists a constant L_k such that for any (x_1, x_2) in $\mathscr{S}^2$, any u in $\mathscr{U}$, and any time $t \geq \overline{t}_{k,u}$

$$|x_1 - x_2| \leq L_k \, |T(x_1, t) - T(x_2, t)| \, .$$

Besides, for any $t \geq \overline{t}_{k,u}$, $T(\cdot, t)$ is an injective immersion on $\mathscr{S}$.

Proof. See Appendix C.1. □

As we will see on examples, it is usually convenient to choose the pair (A_i, B_i) of the form $A_i = -\mathsf{diag}(\lambda_1, \dots, \lambda_{m_i})$ and $B_i = (1, \dots, 1)^\top$ with $m = \max_i m_i$ sufficiently large distinct strictly positive real numbers λ_j. Indeed, the PDEs to solve are then simply

$$\frac{\partial T_{\lambda,i}}{\partial x}(x, t) f(x, u(t)) + \frac{\partial T_{\lambda,i}}{\partial t}(x, t) = -\lambda \, T_{\lambda,i}(x, t) + h_i(x, u(t)) \qquad (6.12)$$

for each $1 \leq i \leq d_y$ and λ in $\{\lambda_1, \dots, \lambda_{m_i}\}$. Then, one takes

$$T(x, t) = \left(T_{\lambda_1,1}, \dots, T_{\lambda_{m_1},1}, \dots, T_{\lambda_1,d_y}, \dots, T_{\lambda_{m_{d_y}},d_y} \right) \, .$$

Note that the additional assumption "$T(\cdot, 0)$ Lipschitz on $\mathcal{S}$" is not very restrictive because the solution T can usually be chosen arbitrarily at initial time 0 (see examples in Sect. 6.3). In particular, the elementary solution T^0 found in Lemma 1 is zero at time 0 and thus clearly verifies this assumption. Actually, this additional assumption is automatically verified when $\mathcal{S}$ is compact, so that the result of Theorem 6.4 holds under the only assumptions of Lemma 6.2 and the point 1 in Assumption 6.2.

Applying successively Lemma 6.1, Theorems 6.4 and 1.1, we conclude that under Assumption 6.2, it is possible to write an observer for system (5.1) by choosing any (A_i, B_i) controllable and k sufficiently large.

6.2.2.3 Injectivity with Backward Distinguishability

In the previous section, we have seen that finding an injective transformation into an Hurwitz form is possible under a strong differential observability property, namely that the function made of each output and a certain number of their derivatives are an injective immersion. It turns out that injectivity can still be ensured under a weak differential observability or even only backward distinguishability. This was proved in [5] and extends [2, Theorem 3] for autonomous systems (recalled in Sect. 6.2.1).

Theorem 6.5 ([5]) *Take u in $\mathcal{U}$. Assume that for this input, System (5.1) is backward distinguishable in time $\bar{t}_u$ on $\mathcal{S}$, i.e., for any $t \geq \bar{t}_u$ and any (x_a, x_b) in $\mathcal{S}^2$,*

$$Y(x_a, t; s; u) = Y(x_b, t; s; u) \quad \forall s \in [t - \bar{t}_u, t] \quad \Longrightarrow \quad x_a = x_b\,.$$

There exists[15] *a set $\mathcal{R}$ of zero Lebesgue measure in $\mathbb{C}^{d_x+1}$ such that for any $(\lambda_1, \ldots, \lambda_{d_x+1})$ in $\Omega^{d_x+1} \setminus \mathcal{R}$ with $\Omega = \{\lambda \in \mathbb{C}, \Re(\lambda) < 0\}$, and any $t \geq \bar{t}_u$, the function T^0 defined in* (6.8) *with*

- $d_\xi = d_y \times (d_x + 1)$
- *A in $\mathbb{R}^{d_\xi \times d_\xi}$ and B in $\mathbb{R}^{d_\xi \times d_y}$ defined by*

$$A = \tilde{A} \otimes I_{d_y}, \quad B = \tilde{B} \otimes I_{d_y}$$

with

$$\tilde{A} = \begin{pmatrix} \lambda_1 & & \\ & \ddots & \\ & & \lambda_{d_x+1} \end{pmatrix}, \quad \tilde{B} = \begin{pmatrix} 1 \\ \vdots \\ 1 \end{pmatrix}.$$

is such that $T^0(\cdot, t)$ is injective on $\mathcal{S}$ for $t > \bar{t}_u$.

Proof. See Appendix C.2. □

[15]This set depends on u, and unfortunately, there is no guarantee that $\bigcup_{u \in \mathcal{U}} \mathcal{R}_u$ is also of zero Lebesgue measure.

Like in Theorem 6.3, the transformation obtained here may take complex values. We refer the reader to Remark 6.1 for details about how to implement it in $\mathbb{R}$.

The assumption of backward distinguishability in finite time is in particular verified when the system is instantaneously backward distinguishable, and a fortiori when the map made of the output and its derivatives up to a certain order is injective; namely, the system is weakly differentially observable.

Of course, if T^0 has been built with the modified dynamics (6.9) instead of (5.1) to satisfy Assumption 6.1, the assumption of backward distinguishability needed here should hold for system (6.9); namely, the outputs should be distinguishable in backward time before the solutions leave $\mathcal{X}$.

Remark 6.2 Unlike Theorem 6.4 which proved the injectivity of any solution T to PDE (6.7), Theorem 6.5 proves only the injectivity of T^0. Note though that under item 1 of Assumption 6.2, any solution T to (6.7) on $\mathcal{S}$ verifies for all s in $\mathbb{R}_{\geq 0}$, and all (x,t) in $\mathcal{S} \times \mathbb{R}_{\geq 0}$,

$$\frac{d}{ds} T(X(x,t;s;u),s) = A\,T(X(x,t;s;u),s) + BY(x,t;s;u)\,.$$

Integrating between times t and s, and taking $s = 0$, T verifies

$$T(x,t) = e^{At}\,T(X(x,t;0;u),0) + T^0(x,t) \tag{6.13}$$

with A Hurwitz. Therefore, as t goes to $+\infty$, $x \mapsto T(x,t)$ "tends" to the injective function $x \mapsto T^0(x,t)$. We can thus expect T to become injective after a certain time under some appropriate uniformity assumptions detailed in [5]. In fact, a way of ensuring injectivity is to take, if possible, a solution T with the boundary condition

$$T(x,0) = 0 \quad \forall x \in \mathcal{S}\,, \tag{6.14}$$

because in that case, necessarily, $T = T^0$ from (6.13). This trick will be used in the examples below.

We conclude from this section that as long as no blowup in finite backward time is possible within $\mathcal{X}$, there always exists a time-varying solution to PDE (6.7) which is injective if the system is backward observable. The computation of such a solution without relying on the expression (6.8) will be shown in the following section through practical examples.

Remark 6.3 It may also be possible to keep a stationary transformation $x \mapsto T(x)$ for controlled systems. For instance, this is the case for a control-affine single-output system that is instantaneously observable for any input[16] and whose drift dynamics are strongly differentially observable of order d_x, as proved in [5, Theorem 4].

[16] Uniformly instantaneously observable, see Definition 1.2.

6.3 Examples

6.3.1 Linear Systems with Unknown Parameters

Consider a linear system

$$\dot{x} = A(\theta)x + B(\theta)u, \quad y = C(\theta)x \tag{6.15}$$

where θ is a vector of unknown constant parameters ranging in a known set Θ, and assume we want to build an observer estimating x and θ. In [1], it is shown how the Luenberger methodology can be efficiently used in that context by adding θ to the state and $\dot{\theta} = 0$ to the dynamics.

Let us try to solve (6.12) for some positive λ. Given the linearity in x of the dynamics, it is tempting to look for $T_{\lambda,i}$ of the form

$$T_{\lambda,i}(x, \theta, t) = M_{\lambda,i}(\theta)x + N_{\lambda,i}(\theta, t)\ .$$

(6.12) is equivalent to

$$M_{\lambda,i}(\theta)A(\theta) = -\lambda M_{\lambda,i}(\theta) + C_i(\theta) \tag{6.16}$$

$$\dot{N}_{\lambda,i}(\theta, t) = -\lambda N_{\lambda,i}(\theta, t) - M_{\lambda,i}(\theta)B(\theta)u\ . \tag{6.17}$$

Now, denoting $\sigma(A(\theta))$ the set of eigenvalues of $A(\theta)$, and if

$$-\lambda \notin \bigcup_{\theta\in\Theta} \sigma(A(\theta))\ , \tag{6.18}$$

(6.16) is equivalent to

$$M_{\lambda,i}(\theta) = C_i(\theta)(A(\theta) + \lambda I_{d_x})^{-1}\ .$$

As for $N_{\lambda,i}$, the differential equation (6.17) cannot be implemented because θ is unknown. Actually, it is straightforward to check that (6.17) holds if

$$N_{\lambda,i}(\theta, t) = -M_{\lambda,i}(\theta)B(\theta)\eta_\lambda(t)$$

with η_λ solution to

$$\dot{\eta}_\lambda = -\lambda\eta_\lambda + u\ . \tag{6.19}$$

In other words, if λ is chosen such that (6.18) holds, a possible solution to (6.12) is

$$T_{\lambda,i}(x, \theta, t) = C_i(\theta)(A(\theta) + \lambda I_{d_x})^{-1}\Big(x - B(\theta)\eta_\lambda(t)\Big)\ , \tag{6.20}$$

with η_λ a filtered version of u solution to (6.19). By definition of (6.12), it means that $\xi_{\lambda,i} = T_{\lambda,i}(x,\theta,t)$ follows the dynamics

$$\dot{\xi}_{\lambda,i} = -\lambda\xi_{\lambda,i} + y_i \ , \tag{6.21}$$

for which a trivial observer is given by

$$\dot{\hat{\xi}}_{\lambda,i} = -\lambda\hat{\xi}_{\lambda,i} + y_i \ , \tag{6.22}$$

Therefore, implementing (6.22) with any initial condition gives an asymptotic estimate of $T_{\lambda,i}(x,\theta,t)$ defined in (6.20).

If the input u confers to the system appropriate observability[17] properties given by Theorem 6.4 or 6.5 and if we take a sufficiently large number of distinct eigenvalues λ for each y_i with i in $\{1,\ldots,d_y\}$, we know that T will become injective after a certain time (once the filters are in steady state). Since T transforms the dynamics into

$$\dot{\xi} = A\xi + B\,y$$

with A Hurwitz, implementing those dynamics for any initial condition gives an estimate of $T(x,\theta,t)$. An estimate for x and θ can therefore be obtained by implementing (6.19) with any initial condition and inverting[18] the map T.

6.3.2 State-Affine Systems with Output Injection and Polynomial Output

Consider a system of the form

$$\dot{x} = A(u,y)\,x + B(u,y), \quad y = C(u)P_d(x) \tag{6.23}$$

with $P_d : \mathbb{R}^{d_x} \to \mathbb{R}^{k_d}$ a vector containing the k_d possible monomials of x with degree inferior to d, and $C : \mathbb{R}^{d_u} \to \mathbb{R}^{d_y \times k_d}$ a matrix of time-varying coefficients. When $d = 1$, a Kalman observer can be designed according to Theorem 3.3 under a regular persistence condition of the extended input (u, y). Suppose this assumption is not verified or $d > 1$. It is observed in [5] that an explicit solution to PDE (6.12) can be found. Indeed, for any i in $\{1,\cdots,d_y\}$, a transformation $T_{\lambda,i}$ of the form

$$T_{\lambda,i}(x,t) = M_{\lambda,i}(t)P_d(x)$$

[17]The conditions given in [1] correspond to those obtained from Theorem 6.4. A weaker condition is given by Theorem 6.5.

[18]In the case where $A(\theta)$ is in companion form, an explicit inversion algorithm is available in [1].

with $M_{\lambda,i} : \mathbb{R} \to \mathbb{R}^{1\times k_d}$, verifies

$$\begin{aligned}\frac{\partial T_{\lambda,i}}{\partial x}(x,t)f(x,u) + \frac{\partial T_{\lambda,i}}{\partial t}(x,t) &= M_{\lambda,i}(t)\frac{\partial P_d}{\partial x}(x)\Big(A(u,y)x + B(u,y)\Big) + \dot{M}_{\lambda,i}(t)P_d(x)\\ &= M_{\lambda,i}(t)D(u,y)P_d(x) + \dot{M}_{\lambda,i}(t)P_d(x)\end{aligned}$$

for a matrix of coefficients $D : \mathbb{R}^{d_u} \times \mathbb{R}^{d_y} \to \mathbb{R}^{k_d}$. It follows that by choosing a positive scalar λ and the coefficients $M_{\lambda,i}$ as solutions of the implementable filters

$$\dot{M}_{\lambda,i} + \lambda\, M_{\lambda,i} = -M_{\lambda,i}(t)D(u,y) + C_i(u)\ , \tag{6.24}$$

$T_{\lambda,i}$ is solution to the PDE (6.12), which, again, means that $\xi_{\lambda,i} = T_{\lambda,i}(x,t)$ follows the dynamics (6.21) for which a trivial observer is given by (6.22). Therefore, implementing (6.24)–(6.22) with any initial condition for i in $\{1,\ldots,d_y\}$ gives an asymptotic estimate of $T_{\lambda,i}(x,t)$. Besides, if the observability conditions given by Theorem 6.4 or 6.5 are satisfied, one obtain injectivity of the transformation after a finite time by taking a sufficiently large number of distinct λ.

The reader may have observed that taking the transformation made of the monomials $T = P_d$ also enables to obtain a linear normal form with time-varying linear part. Indeed, the dynamics being linear, $\dot{P}_d$ is still a linear combination of monomials of degree inferior to d. Therefore, a Kalman observer could be used but the problem is to verify the regular persistent excitation condition of Theorem 3.3 for the new system in the ξ-coordinates. Indeed, even if the initial system is observable, the linear normal form, which is of significant higher dimension, may not be.

A practical example presented in [4, 18] is a permanent magnet synchronous motor (PMSM), which can be modeled by

$$\dot{x} = u - Ri \quad , \quad y = |x - Li|^2 - \Phi^2 = 0 \tag{6.25}$$

where x is in $\mathbb{R}^2$, the voltages u and currents i are inputs in $\mathbb{R}^2$, the resistance R, impedance L, and flux Φ are known scalar parameters, and the measurement y is constantly zero. It can be checked that if i, $\dot{\widehat{i}}$, $\ddot{\widehat{i}}$ and u, $\dot{u}$ are bounded, the state x also remains bounded. Choosing $m = 3$, the function H defined in Assumption 6.2 is given by ($d_y = 1$)

$$H(x,u,i,\dot{u},\dot{\widehat{i}},\ddot{\widehat{i}}) = \begin{pmatrix} |x - Li|^2 - \Phi^2 \\ 2\eta^\top(x - Li) \\ 2\dot{\eta}^\top(x - Li) + 2\eta^\top\eta \end{pmatrix}$$

where we denote $\eta = u - Ri + L\dot{\widehat{i}}$. Because the inputs happen to be such that[19] $\det(\eta\ ,\ \dot{\eta}) = \omega^3\Phi^2$, where ω is the rotor angular velocity, every assumption of Lemma 6.2 is satisfied when the inputs and their derivatives are bounded and the

[19] $\eta = \Phi\omega\begin{pmatrix} -\sin\theta \\ \cos\theta \end{pmatrix}$, with θ the electrical angle, and $\omega = \dot{\theta}$.

rotor angular velocity is away from zero. Applying Theorem 6.4, it follows that for any three distinct and sufficiently large strictly positive λ_j, the function

$$T(x,t) = (T_{\lambda_1}(x,t), T_{\lambda_2}(x,t), T_{\lambda_3}(x,t))$$

becomes injective after a certain time. Implementing (6.24)–(6.22) for each λ_j, one can obtain after a certain time an estimate $\hat{x}$ of $x(t)$ by inverting the map T. An explicit expression of T and of its inverse is given in [4].

6.3.3 Non-holonomic Vehicle

Another appropriate example is the non-holonomic vehicle with dynamics

$$\begin{cases} \dot{x}_1 = u_1 \cos(x_3) \\ \dot{x}_2 = u_1 \sin(x_3) \\ \dot{x}_3 = u_1 u_2 \end{cases}, \quad y = (x_1, x_2) \tag{6.26}$$

where the inputs u_1 and u_2 correspond to the norm of vehicle velocity and the orientation of the front steering wheels, respectively.

Taking $T(x) = (x_1, x_2, \cos(x_3), \sin(x_3))$ enables to linearize the system since we obtain

$$\begin{cases} \dot{\xi}_1 = u_1 \xi_3 \\ \dot{\xi}_2 = u_1 \xi_4 \\ \dot{\xi}_3 = -u_1 u_2 \xi_4 \\ \dot{\xi}_4 = u_1 u_2 \xi_3 \end{cases}, \quad y = (\xi_1, \xi_2)$$

and a Kalman observer can be used provided the excitation condition of Theorem 3.3 is satisfied. As mentioned above, this is not guaranteed by the observability of the initial system. However, in this particular case, computing the successive derivatives of the measurements (ξ_1, ξ_2) enables to see that the linearized system is observable if at each time, u_1 or one of its derivatives is nonzero. The persistence of excitation condition is satisfied if this is verified uniformly in time. This gives an observer of dimension[20] $d_\xi + \dim P = d_\xi + \frac{d_\xi(d_\xi+1)}{2} = 14$.

As noticed in [4], a nonlinear Luenberger design is also possible. The dynamics and measurements being linear in $x_1, x_2, \cos(x_3), \sin(x_3)$, it is quite natural to look for a function T linear in those quantities. Since x_1 and x_2 are independent, we take $T_{\lambda,1}$ and $T_{\lambda,2}$ of the form

$$\begin{aligned} T_{\lambda,1}(x,t) &= a_\lambda(t)\, x_1 + b_\lambda(t) \cos(x_3) + c_\lambda(t) \sin(x_3) \\ T_{\lambda,2}(x,t) &= \tilde{a}_\lambda(t)\, x_2 + \tilde{b}_\lambda(t) \cos(x_3) + \tilde{c}_\lambda(t) \sin(x_3)\,. \end{aligned}$$

[20] P is symmetric so it is sufficient to compute $\frac{d_\xi(d_\xi+1)}{2}$ components.

Straightforward computations show that the choice

$$\begin{aligned}\tilde{a}_\lambda = a_\lambda = \frac{1}{\lambda} \quad , \quad \tilde{b}_\lambda = -c_\lambda, \quad \tilde{c}_\lambda = b_\lambda \\ \dot{b}_\lambda = -\lambda\, b_\lambda - u_1 u_2\, c_\lambda - \frac{1}{\lambda} u_1 \\ \dot{c}_\lambda = -\lambda\, c_\lambda + u_1 u_2\, b_\lambda\,. \end{aligned} \tag{6.27}$$

ensures that $T_{\lambda,1}$ and $T_{\lambda,2}$ are solutions of

$$\begin{aligned}\dot{\xi}_{\lambda,1} &= -\lambda\, \xi_{\lambda,1} + x_1 \\ \dot{\xi}_{\lambda,2} &= -\lambda\, \xi_{\lambda,2} + x_2\end{aligned} \tag{6.28}$$

respectively. Besides, the computation of the successive derivatives of the measurements (x_1, x_2) shows that H_1 and H_2 are injective immersions at the order m if at least u_1 or one of its first $m-2$ derivatives is nonzero. Therefore, if the state, the inputs u_1, u_2, and their derivatives remain in compact sets, and if $u_1(t)^2 + \dot{u}_1(t)^2 + \cdots + u_1^{(m-2)}(t)^2$ remains uniformly away from 0, then all the assumptions in Lemma 6.2 are verified with $m_1 = m_2 = m$. Therefore, by choosing m strictly positive distinct real numbers λ_j, the function

$$T(x,t) = (T_{\lambda_1,1}(x,t), \ldots, T_{\lambda_m,1}(x,t), T_{\lambda_1,2}(x,t), \ldots, T_{\lambda_m,2}(x,t))$$

becomes injective after a certain time. Implementing (6.27)–(6.28) for each λ_j, we thus get an observer of dimension $4m$. Actually, since the system is instantaneously backward distinguishable as long as u_1 does not stay at zero, we also know from Theorem 6.5 that for almost all $d_x + 1 = 4$ complex λ_j with positive real parts, T should become injective after a certain time. So, in the worst case scenario, we obtain an observer of dimension 16.

6.3.4 *Time-Varying Transformations for Autonomous Systems*

As in [5], consider, for instance, the system

$$\begin{cases} \dot{x}_1 = x_2^3 \\ \dot{x}_2 = -x_1 \end{cases}, \quad y = x_1 \tag{6.29}$$

which admits bounded trajectories, the quantity $x_1^2 + x_2^4$ being constant along the trajectories. This system is weakly differentially observable of order 2 on $\mathbb{R}^2$ since $x \mapsto \mathbf{H}_2(x) = (x_1, x_2^3)$ is injective on $\mathbb{R}^2$. It is thus a fortiori instantaneously backward distinguishable, and Theorem 6.3 holds. However, we are not able to compute explicitly an expression of the stationary transformation (6.6) in that case.

Of course, we could also linearize the system by taking the map $T(x) = (x_1, x_2^3, x_2^2, x_2)$, and obtain

$$\begin{cases} \dot{\xi}_1 = \xi_2 \\ \dot{\xi}_2 = -3y\xi_3 \\ \dot{\xi}_3 = -2y\xi_4 \\ \dot{\xi}_4 = -y \end{cases}, \quad y = \xi_1$$

But this system is observable only if y stays "regularly" (and uniformly) away from 0. If this is the case, we obtain a Kalman observer of dimension $d_\xi + \frac{d_\xi(d_\xi+1)}{2} = 14$.

Let us try the Luenberger route with a time-varying transformation. Following [5], given the structure of the dynamics and the previous linearization, we look for a transformation of the form

$$T_\lambda(x,t) = a_\lambda(t)x_2^3 + b_\lambda(t)x_2^2 + c_\lambda(t)x_2 + d_\lambda(t)x_1 + e_\lambda(t) \,. \tag{6.30}$$

It verifies the dynamics

$$\dot{\xi}_\lambda = -\lambda\,\xi_\lambda + x_1 \,, \tag{6.31}$$

if, for instance,

$$\begin{aligned} \dot{a}_\lambda &= -\lambda\, a_\lambda - d_\lambda \\ \dot{b}_\lambda &= -\lambda\, b_\lambda + 3a_\lambda y \\ \dot{c}_\lambda &= -\lambda\, c_\lambda + 2b_\lambda y \\ \dot{d}_\lambda &= -\lambda\, d_\lambda + 1 \\ \dot{e}_\lambda &= -\lambda\, e_\lambda + c_\lambda y \end{aligned}$$

Using Remark 6.2 and applying Theorem 6.5, $x \mapsto (T_{\lambda_1}(x,t), T_{\lambda_2}(x,t), T_{\lambda_3}(x,t))$ is injective on $\mathbb{R}^2$ for $t \geq 0$ and for a generic choice of $(\lambda_1, \lambda_2, \lambda_3)$ in $\{\lambda \in \mathbb{C} : \Re(\lambda) > 0\}^3$, if the filters are initialized at 0 at time 0. Actually, to reduce the observer dimension, we can choose $d_\lambda(t) = \frac{1}{\lambda}$ and $a_\lambda(t) = \frac{1}{\lambda^2}$. In that case, the use of Theorem 6.5 is no longer direct because T_λ is not T_λ^0. However, according to Remark 6.2, and because the trajectories evolve in a compact set, injectivity should still be ensured after a certain time. Finally, to recover $\hat{x}$ from $\hat{\xi}$, T must be inverted: This can be done by first linearly combining the $T_{\lambda_i} - \xi_{\lambda_i}$ to make x_1 disappear (because the d_{λ_i} are all nonzero) and then searching numerically the common roots of the obtained two polynomials of order 3 in x_2. With (6.31) and the filters for the transformations, this gives an observer of dimension 12, without any assumptions on the trajectories.

References

1. Afri, C., Andrieu, V., Bako, L., Dufour, P.: State and parameter estimation: a nonlinear Luenberger approach (2015). arXiv preprint arXiv:1511.07687

2. Andrieu, V., Praly, L.: On the existence of a Kazantzis-Kravaris/Luenberger observer. SIAM J. Control Optim. **45**(2), 432–456 (2006)
3. Back, J., Seo, J.: Immersion of non-linear systems into linear systems up to output injection: characteristic equation approach. Int. J. Control **77**(8), 723–734 (2004)
4. Bernard, P.: Luenberger observers for nonlinear controlled systems. In: IEEE Conference on Decision and Control (2017)
5. Bernard, P., Andrieu, V.: Luenberger observers for non autonomous nonlinear systems. IEEE Trans. Autom. Control **64**(1), 270–281 (2019)
6. Besançon, G., Bornard, G.: On characterizing a class of observer forms for nonlinear systems. In: European Control Conference (1997)
7. Bestle, D., Zeitz, M.: Canonical form observer design for nonlinear time variable systems. Int. J. Control. **38**, 419–431 (1983)
8. Bossane, D., Rakotopara, D., Gauthier, J.P.: Local and global immersion into linear systems up to output injection. In: IEEE Conference on Decision and Control, pp. 2000–2004 (1989)
9. Engel, R.: Exponential observers for nonlinear systems with inputs. Universität of Kassel, Department of Electrical Engineering, Technical report (2005). https://doi.org/10.13140/RG.2.2.34476.05764
10. Engel, R.: Nonlinear observers for Lipschitz continuous systems with inputs. Int. J. Control **80**(4), 495–508 (2007)
11. Fliess, M.: Finite-dimensional observation-spaces for non-linear systems. In: Hinrichsen, D., Isidori, A. (eds.) Feedback Control of Linear and Nonlinear Systems. Lecture Notes in Control and Information Sciences, vol. 39, pp. 73–77. Springer, Berlin (1982)
12. Fliess, M., Kupka, I.: A finiteness criterion for nonlinear input-output differential systems. SIAM J. Control. Optim. **21**(5), 721–728 (1983)
13. Glumineau, A., Moog, C.H., Plestan, F.: New algebro-geometric conditions for the linearization by input-output injection. IEEE Trans. Autom. Control **41**(4), 598–603 (1996)
14. Guay, M.: Observer linearization by output-dependent time-scale transformations. IEEE Trans. Autom. Control **47**(10), 1730–1735 (2002)
15. Hamami, Y.: Observateur de Kazantzis-Kravaris/Luenberger dans le cas d'un système instationnaire. Technical report, MINES ParisTech (2008)
16. Hammouri, H., Celle, F.: Some results about nonlinear systems equivalence for the observer synthesis. In: Trends in Systems Theory. New Trends in Systems Theory, vol. 7, pp. 332–339. Birkhäuser (1991)
17. Hammouri, H., Kinnaert, M.: A new procedure for time-varying linearization up to output injection. Syst. Control Lett. **28**, 151–157 (1996)
18. Henwood, N., Malaizé, J., Praly, L.: A robust nonlinear Luenberger observer for the sensorless control of SM-PMSM: rotor position and magnets flux estimation. In: IECON Conference on IEEE Industrial Electronics Society (2012)
19. Jouan, P.: Immersion of nonlinear systems into linear systems modulo output injection. SIAM J. Control Optim. **41**(6), 1756–1778 (2003)
20. Kazantzis, N., Kravaris, C.: Nonlinear observer design using Lyapunov's auxiliary theorem. Syst. Control Lett. **34**, 241–247 (1998)
21. Keller, H.: Nonlinear observer by transformation into a generalized observer canonical form. Int. J. Control **46**(6), 1915–1930 (1987)
22. Kreisselmeier, G., Engel, R.: Nonlinear observers for autonomous lipshitz continuous systems. IEEE Trans. Autom. Control **48**(3), 451–464 (2003)
23. Krener, A., Isidori, A.: Linearization by output injection and nonlinear observers. Syst. Control Lett. **3**, 47–52 (1983)
24. Krener, A., Xiao, M.: Nonlinear observer design in the Siegel domain. SIAM J. Control Optim. **41**(3), 932–953 (2003)
25. Krener, A., Xiao, M.: Nonlinear observer design for smooth systems. In: Perruguetti, W., Barbot., J.-P. (eds.) Chaos in Automatic Control, pp. 411–422. Taylor and Francis, Routledge (2006)

26. Krener, A.J., Respondek, W.: Nonlinear observers with linearizable dynamics. SIAM J. Control Optim. **23**(2), 197–216 (1985)
27. Luenberger, D.: Observing the state of a linear system. IEEE Trans. Mil. Electron. **8**, 74–80 (1964)
28. Plestan, F., Glumineau, A.: Linearization by generalized input-output injection. Syst. Control Lett. **31**, 115–128 (1997)
29. Respondek, W., Pogromski, A., Nijmeijer, H.: Time scaling for observer design with linearizable error dynamics. Automatica **40**, 277–285 (2004)
30. Rotella, F., Zambettakis, I.: On functional observers for linear time-varying systems. IEEE Trans. Autom. Control **58**(5), 1354–1360 (2013)
31. Shoshitaishvili, A.: On control branching systems with degenerate linearization. In: IFAC Symposium on Nonlinear Control Systems, pp. 495–500 (1992)
32. Trumpf, J.: Observers for linear time-varying systems. Linear Algebr. Appl. **425**, 303–312 (2007)

Chapter 7
Transformation Into Triangular Forms

The basic idea to transform a system into triangular dynamics is to consider a transformation made of the output map and a certain number of its derivatives. It has been known since [3] that any uniformly instantaneously observable[1] single-output control-affine system, whose drift system is strongly differentially observable[2] of order its dimension d_x, can be transformed into a Lipschitz triangular form (4.2). In the more general case where the order of differential observability is larger than the dimension of the system, it was recently shown in [1] how the system dynamics may still be described by a continuous triangular form but with nonlinear functions Φ_i which may not be locally Lipschitz.

7.1 Lipschitz Triangular Form

The Lipschitz triangular form[3] (4.2)

$$\begin{cases} \dot{\xi}_1 = \xi_2 + \Phi_1(\tilde{u}, \xi_1) \\ \quad \vdots \\ \dot{\xi}_i = \xi_{i+1} + \Phi_i(\tilde{u}, \xi_1, \dots, \xi_i) \\ \quad \vdots \\ \dot{\xi}_m = \Phi_m(\tilde{u}, \xi) \end{cases} , \quad y = \xi_1$$

[1]See Definition 1.2.

[2]See Definition 5.3.

[3]It is useful here to denote the input $\tilde{u}$ instead of u because we will see that Φ can sometimes depend on $\tilde{u} = (u, \dot{u}, \ddot{u}, \dots)$.

P. Bernard, *Observer Design for Nonlinear Systems*,
Lecture Notes in Control and Information Sciences 479,
https://doi.org/10.1007/978-3-030-11146-5_7

is well known because it is associated with the classical high-gain observer (4.4). The idea of transforming a nonlinear system into a phase-variable form[4] (i.e., with $\Phi_i = 0$ except Φ_m) appeared in [13]. For an autonomous system,

$$\dot{x} = f(x) \quad , \quad y = h(x)$$

the function $\mathbf{H}_m$ defined by the output and its $m-1$ first derivatives, namely

$$\mathbf{H}_m(x) = (h(x), L_f h(x), \dots, L_f^{m-1} h(x)) \ ,$$

transforms the system into

$$\dot{\xi}_1 = \xi_2 \quad , \quad \dots \quad , \quad \dot{\xi}_i = \xi_{i+1} \quad , \quad \dots \quad , \quad \dot{\xi}_m = L_f^m h(x) \quad , \quad y = \xi_1 \ .$$

This is a Lipschitz phase-variable form if and only if there exists a function Φ_m Lipschitz on $\mathbb{R}^{d_\xi}$ such that

$$\forall x \in \mathcal{X} \quad , \quad L_f^m h(x) = \Phi_m(\mathbf{H}_m(x)) \ ,$$

i.e., the mth derivative of the output can be expressed "in a Lipschitz way" in terms of its $m-1$ first derivatives. This is possible, for example, if $\mathcal{X}$ is bounded and $\mathbf{H}_m$ is an injective immersion[5] on some open set $\mathcal{S}$ containing $\texttt{cl}(\mathcal{X})$ (see Theorem 7.1 below for this result in the general controlled case). In the remaining of this section, we review the existing results in terms of transformation of general controlled systems into a Lipschitz triangular form.

7.1.1 Time-Varying Transformation

The first natural idea introduced in [13] is to keep considering the transformation made of the output and its $m-1$ first derivatives, despite the presence of the input, and transform the system into a phase-variable form in the same way as for autonomous systems. Recall that we have defined in Definition 5.2 a map $\overline{\mathbf{H}}_m(\cdot, \overline{\nu}_m)$ made of the successive derivatives of the output, thanks to the dynamic extension defined in Definition 5.1. The function $\overline{\mathbf{H}}_m(\cdot, \overline{\nu}_m)$ is equivalent to $\mathbf{H}_m$ for autonomous systems, but it now depends on the input and its derivatives. A straightforward extension of the stationary case along the idea presented in [13] is therefore the following.

Theorem 7.1 *If $\overline{U}_m$ is a compact subset of $\mathbb{R}^{d_u(m+1)}$ and there exists an integer m and a subset $\mathcal{S}$ of $\mathbb{R}^{d_x}$ such that System* (5.1) *is weakly (resp strongly) differentially observable of order m on $\mathcal{S}$, then, for any compact subset $\mathcal{C}$ of $\mathcal{S}$ and any u in $\mathcal{U}$,*

---

[4]See Definition 4.1.

[5]$\mathbf{H}_m$ is injective, and $\frac{\partial \mathbf{H}_m}{\partial x}(x)$ has full-rank on $\mathcal{X}$.

the function defined by

$$T(x,t) = \overline{\mathbf{H}}_m(x, \overline{u}_m(t))$$

transforms System (5.1) *into a continuous (resp Lipschitz) phase-variable form of dimension* $d_\xi = m d_y$ *on* $\mathcal{C}$ *and with input* $\tilde{u} = \overline{u}_m$. *Besides,* $x \mapsto T(x,t)$ *is uniformly injective in space and in time on* $\mathcal{C}$.

Proof Assume first that the system is weakly differentially observable of order m; i.e., for all $\overline{\nu}_m$ in $\overline{U}_m$, $x \mapsto \overline{\mathbf{H}}_m(x, \overline{\nu}_m)$ is injective on $\mathcal{C}$. According to Lemma A.12, it is uniformly injective in space and in time on $\mathcal{C}$ and for any $\overline{\nu}_m$, it admits a uniformly continuous left inverse; i.e., there exists a function $\overline{\mathbf{H}}_m^{-1} : \mathbb{R}^{d_\xi} \times \mathbb{R}^{d_u(m+1)} \to \mathbb{R}^{d_x}$ such that for all $\overline{\nu}_m$ in $\overline{U}_m$ and all x in $\mathcal{C}$

$$x = \overline{\mathbf{H}}_m^{-1}(\overline{\mathbf{H}}_m(x, \overline{\nu}_m), \overline{\nu}_m) \ .$$

Now, define

$$\Phi_m(\xi, \overline{\nu}_m) = L_{\overline{f}}^m \bar{h}(\overline{\mathbf{H}}_m^{-1}(\xi, \overline{\nu}_m), \overline{\nu}_m) \ .$$

T transforms System (5.1) into the continuous phase-variable form

$$\begin{cases} \dot{\xi}_1 = \xi_2 \\ \quad\vdots \\ \dot{\xi}_{m-1} = \xi_m \\ \dot{\xi}_m = \Phi_m(\xi, \overline{u}_m(t)) \end{cases}$$

Assume now the system is strongly observable. Still with Lemma A.12, $\xi \mapsto \overline{\mathbf{H}}_m^{-1}(\xi, \overline{\nu}_m)$ can now be taken Lipschitz on $\mathbb{R}^{d_\xi}$, with the same Lipschitz constant for all $\overline{\nu}_m$ in $\overline{U}_m$. It follows that $\xi \mapsto \Phi_m(\xi, \overline{\nu}_m)$ is Lipschitz on any compact set of $\mathbb{R}^{d_\xi}$ containing the compact set of interest $\overline{\mathbf{H}}_m(\mathcal{C} \times \overline{U}_m)$, with the same Lipschitz constant for all $\overline{\nu}_m$ in $\overline{U}_m$. According to Kirszbraun–Valentine theorem [8, 12], it can be extended to a Lipschitz function on $\mathbb{R}^{d_\xi}$ with still the same Lipschitz constant. This new extended function $\xi \mapsto \Phi_m(\xi, \overline{\nu}_m)$ is globally Lipschitz uniformly in $\overline{\nu}_m$ and has not changed on $\overline{\mathbf{H}}_m(\mathcal{C} \times \overline{U}_m)$ where the system solutions evolve; thus, we have a Lipschitz phase-variable form. □

The assumptions given in Theorem 7.1 are sufficient to ensure the existence of the function Φ_m in the phase-variable form. But they are not necessary. The possibility of finding such a function, namely to express $L_{\overline{f}}^m \bar{h}$ (the mth derivative of the output) in terms of h, $L_{\overline{f}} \bar{h}, \ldots, L_{\overline{f}}^{m-1} \bar{h}$ (the output and its $m-1$ first derivatives), and $\overline{u}_m$ (the input and its m first derivatives), is thoroughly studied in [7] through the so-called ACP(m) condition. We refer the reader to [7] (or [5]) for a more complete analysis of those matters.

Remark 7.1 Note that as we saw in Sect. 4.1.5 of Chap. 4, to design a high-gain observer, it is not necessary to have global Lipschitzness of the function Φ_m with respect to ξ. It is sufficient to have

$$|\Phi_m(\xi, \overline{v}_m) - \Phi_m(\hat{\xi}, \overline{v}_m)| \leq \mathfrak{a}|\xi - \hat{\xi}|$$

for all $\hat{\xi}$ in $\mathbb{R}^{d_\xi}$, all $\overline{v}_m$ in $\overline{U}_m$ and ξ in a compact set containing $\overline{\mathbf{H}}_m(\mathscr{C} \times \overline{U}_m)$ where the system solutions evolve. Thus, the Lipschitz extensions made in the proof of Theorem 7.1 are not necessary in practice and it is sufficient to take[6]

$$\Phi_m(\xi, \overline{v}_m) = \mathsf{sat}_M(L_f^m \hbar(\overline{\mathbf{H}}_m^{-1}(\xi, \overline{v}_m), \overline{v}_m)) \tag{7.1}$$

where M is a bound for $|L_f^m \hbar|$ on $\mathscr{C} \times \overline{U}_m$ and $\overline{\mathbf{H}}_m^{-1}$ is any locally Lipschitz function defined on $\mathbb{R}^{d_\xi} \times \overline{U}_m$ which is a left inverse of $\overline{\mathbf{H}}_m$ on $\overline{\mathbf{H}}_m(\mathscr{C} \times \overline{U}_m)$. It follows that the only difficulty is the computation of a globally defined left inverse for $\overline{\mathbf{H}}_m$, which is needed anyway to deduce an estimate $\hat{x}$ from $\hat{\xi}$ (see [11]).

7.1.2 *Stationary Transformation*

We have seen that under an appropriate injectivity assumption, the function made of the successive derivatives of the output transforms the system into a Lipschitz phase-variable form. The drawback of this design is that the transformation depends on the derivatives of the input, which we may not have access to, in particular if we are not in an output feedback configuration. It turns out that under the assumptions of uniform instantaneous observability and strong differential observability of the drift system of order d_x, a control-affine multi-input single-output system

$$\dot{x} = f(x) + g(x)u \quad , \quad y = h(x) \in \mathbb{R} \tag{7.2}$$

can be transformed into a Lipschitz triangular form (4.2) by a stationary transformation. This famous result was first proved in [3] and then in a simpler way in [4].

Theorem 7.2 ([3, 4]) *Assume that there exists an open subset $\mathscr{S}$ of $\mathbb{R}^{d_x}$ such that*

- *System* (7.2) *is uniformly instantaneously observable*[7] *on $\mathscr{S}$.*
- *The drift system of System* (7.2) *is strongly differentially observable*[8] *of order d_x on $\mathscr{S}$.*

Then, $\mathbf{H}_{d_x}$ defined by

$$\mathbf{H}_{d_x}(x) = (h(x), L_f h(x), \ldots, L_f^{d_x - 1} h(x)), \tag{7.3}$$

[6] The saturation function is defined by $\mathsf{sat}_M(s) = \min\{M, \max\{s, -M\}\}$.

[7] See Definition 1.2.

[8] See Definition 5.3.

which is a diffeomorphism on $\mathcal{S}$ by assumption, transforms System (7.2) *into a Lipschitz triangular form* (4.2) *of dimension $d_\xi = d_x$ on $\mathcal{S}$.*

Proof Consequence from the more general Theorem 7.6 resulting from the results proved in Sect. 7.2. □

Triangularity makes the form (4.2) instantaneously observable for any input. Since the transformation $\mathbf{H}_{d_x}$ itself is independent from the input and injective, this observability property must necessarily be verified by the original System (7.2). Thus, the first assumption is necessary. A usual case where this property is verified is when there exists an order p such that the system is weakly differentially observable[9] of order p.

It is crucial that the order of strong differential observability of the drift system is d_x (the dimension of the state) to ensure the Lipschitzness of the triangular form in order to use a high-gain observer. When this order is larger than d_x, we will see in the next section that triangularity is often preserved but the Lipschitzness is lost.

7.2 Continuous Triangular Form

We carry[10] on considering a single-output control-affine system (7.2), but we would like to relax the condition of strong differential observability of order d_x and see if a triangular form can still be obtained at least up to a certain order and on compact sets.

Problem 7.1 *(Up-to-$\daleth$-triangular form)* Given a compact subset $\mathcal{C}$ of $\mathbb{R}^{d_x}$, under which condition do there exist integers $\daleth$ and d_ξ, a continuous injective function $T : \mathcal{C} \to \mathbb{R}^{d_\xi}$, and continuous functions $\varphi_{d_\xi} : \mathbb{R}^{d_\xi} \to \mathbb{R}$ and $\mathfrak{g}_i : \mathbb{R}^i (\text{or } \mathbb{R}^{d_\xi}) \to \mathbb{R}^{d_u}$ such that T transforms System (7.2) into the up-to-$\daleth$-triangular form

$$\left\{ \begin{aligned} \dot{\xi}_1 &= \xi_2 + \mathfrak{g}_1(\xi_1)\, u \\ &\;\vdots \\ \dot{\xi}_\daleth &= \xi_{\daleth+1} + \mathfrak{g}_\daleth(\xi_1, \ldots, \xi_\daleth)\, u \\ \dot{\xi}_{\daleth+1} &= \xi_{\daleth+2} + \mathfrak{g}_{\daleth+1}(\xi)\, u \\ &\;\vdots \\ \dot{\xi}_{d_\xi} &= \varphi_{d_\xi}(\xi) + \mathfrak{g}_{d_\xi}(\xi)\, u \end{aligned} \right. \quad , \quad y = x_1 \tag{7.4}$$

on $\mathcal{C}$.

Because $\mathfrak{g}_i$ depends only on ξ_1 to ξ_i, for $i \leq \daleth$, but potentially on all the components of ξ for $i > \daleth$, we call this particular form *up-to-$\daleth$-triangular normal form*

[9] See Definition 5.2.

[10] Texts of Sect. 7.2 are reproduced from [1] with permission from Elsevier.

and $\mathfrak{T}$ is called the order of triangularity. When $d_\xi = \mathfrak{T}+1$, we say full triangular normal form. When the functions φ_{d_ξ} and $\mathfrak{g}_j$ are locally Lipschitz, we say Lipschitz up-to-$\mathfrak{T}$-triangular normal form.

We have just seen with Theorem 7.2 that if System (5.1) is instantaneously uniformly observable and $\mathbf{H}_{d_x}$ is a diffeomorphism on an open set $\mathcal{S}$ containing the given compact set $\mathcal{C}$, $T = \mathbf{H}_{d_x}$ transforms the system on $\mathcal{C}$ into a full Lipschitz triangular normal form of dimension $d_\xi = d_x$. However, in general, it can happen that the system is not strongly differentially observable of order d_x everywhere. This motivates our interest in the case where the drift system is strongly differentially observable of order $m > d_x$ (i.e., $\mathbf{H}_m$ is an injective immersion but not a diffeomorphism), or even in the case where the drift system is only weakly differentially observable of order m (i.e., $\mathbf{H}_m$ is injective).

The specificity of the triangular normal form (7.4) is not so much in its structure but more in the dependence of its functions $\mathfrak{g}_i$ and φ_{d_ξ}. Indeed, by choosing $T = \mathbf{H}_{d_\xi}$, we obtain in general:

$$\dot{\widehat{\mathbf{H}_{d_\xi}(x)}} = \begin{pmatrix} 0 & 1 & 0 & \dots & 0 \\ \vdots & \ddots & \ddots & \ddots & \vdots \\ \vdots & & \ddots & \ddots & 0 \\ 0 & \dots & \dots & 0 & 1 \\ 0 & \dots & \dots & \dots & 0 \end{pmatrix} \mathbf{H}_{d_\xi}(x) + \begin{pmatrix} 0 \\ \vdots \\ \vdots \\ 0 \\ L_f^{d_\xi} h(x) \end{pmatrix} + L_g \mathbf{H}_{d_\xi}(x) u$$

But, to get (7.4), we need further the existence of functions φ_{d_ξ} and $\mathfrak{g}_i$ satisfying, for $i > \mathfrak{T}$,

$$L_f^{d_\xi} h(x) = \varphi_{d_\xi}(\mathbf{H}_{d_\xi}(x)) \quad , \quad L_g L_f^{i-1}(x) = \mathfrak{g}_i(\mathbf{H}_{d_\xi}(x)) \quad \forall x \in \mathcal{C} \tag{7.5}$$

and, for $i \leq \mathfrak{T}$,

$$L_g L_f^{i-1}(x) = \mathfrak{g}_i(h(x), \dots, L_f^{i-1} h(x)) \quad \forall x \in \mathcal{C}\,. \tag{7.6}$$

Let us illustrate via the following elementary example what can occur.

Example 7.1 Consider the system defined as

$$\begin{cases} \dot{x}_1 = x_2 \\ \dot{x}_2 = x_3^3 \\ \dot{x}_3 = 1 + u \end{cases} \quad , \quad y = x_1$$

We get

$$\mathbf{H}_3(x) = \begin{pmatrix} h(x) \\ L_f h(x) \\ L_f^2 h(x) \end{pmatrix} = \begin{pmatrix} x_1 \\ x_2 \\ x_3^3 \end{pmatrix} \quad , \quad \mathbf{H}_5(x) = \begin{pmatrix} \mathbf{H}_3(x) \\ L_f^3 h(x) \\ L_f^4 h(x) \end{pmatrix} = \begin{pmatrix} \mathbf{H}_3(x) \\ 3x_3^2 \\ 6x_3 \end{pmatrix}$$

Hence, $\mathbf{H}_3$ is a bijection and $\mathbf{H}_5$ is an injective immersion on $\mathbb{R}^3$. So, the drift system is weakly differentially observable of order 3 on $\mathbb{R}^3$ and strongly differentially observable of order 5 on $\mathbb{R}^3$. Also, the function $(x_1, x_2, x_3) \mapsto (y, \dot{y}, \ddot{y})$ being injective for all u, it is uniformly instantaneously observable on $\mathbb{R}^3$. From this, we could be tempted to pick $d_\xi = 3$ or 5 and the compact set $\mathscr{C}$ arbitrary in $\mathbb{R}^3$. Unfortunately, if we choose $d_\xi = 3$, we must have

$$\varphi_3(\mathbf{H}_3(x)) = L_f^3 h(x) \;=\; 3x_3^2 \;=\; 3(L_f^2 h(x))^{2/3}$$

and there is no locally Lipschitz function φ_3 satisfying (7.5) if the given compact set $\mathscr{C}$ contains a point satisfying $x_3 = 0$. If we choose $d_\xi = 5$, we must have

$$\mathfrak{g}_3(\mathbf{H}_3(x)) = L_g L_f^2 h(x) = 3x_3^2 = L_f^3 h(x) = 3(L_f^2 h(x))^{2/3}$$

and there is no locally Lipschitz function $\mathfrak{g}_3$ satisfying (7.6) if the given compact set $\mathscr{C}$ contains a point satisfying $x_3 = 0$. ▲

Following this example, we leave aside the Lipschitzness requirement for the time being and focus on the existence of continuous functions φ_{d_ξ} and $\mathfrak{g}_i$ verifying (7.5) and (7.6). It turns out that (7.5) is easily satisfied as soon as $\mathbf{H}_{d_\xi}$ is injective.

Theorem 7.3 ([1]) *Suppose the drift system of System (5.1) is weakly (resp. strongly) differentially observable of order m on an open set $\mathscr{S}$ containing the given compact set $\mathscr{C}$. For any $d_\xi \geq m$, there exist continuous (resp. Lipschitz) functions $\varphi_{d_\xi} : \mathbb{R}^{d_\xi} \to \mathbb{R}$, $\mathfrak{g}_i : \mathbb{R}^{d_\xi} \to \mathbb{R}$ satisfying (7.5).*

Proof It is a direct consequence from the fact that a continuous injective function, like $\mathbf{H}_m$, defined on a compact set admits a continuous left inverse defined on $\mathbb{R}^{d_\xi}$ (see Lemma A.10), and that when it is also an immersion, its left inverse can be chosen Lipschitz on $\mathbb{R}^{d_\xi}$ (see Lemma A.12 or [11]). □

We conclude that the real difficulty lies in finding triangular functions $\mathfrak{g}_i$ satisfying (7.6).

7.2.1 Existence of $\mathfrak{g}_i$ Satisfying (7.6)

7.2.1.1 Main Result

The following result was proved in [1].

Theorem 7.4 ([1]) *Suppose System (5.1) is uniformly instantaneously observable on an open set $\mathscr{S}$ containing the given compact set $\mathscr{C}$. Then,*

- *There exists a continuous function $\mathfrak{g}_1 : \mathbb{R} \to \mathbb{R}^{d_u}$ satisfying (7.6).*
- *If, for some i in $\{2, \ldots, d_x\}$, $\mathbf{H}_2, \ldots, \mathbf{H}_i$ defined in* (5.6) *are open maps, then, for all $j \leq i$, there exists a continuous function $\mathfrak{g}_j : \mathbb{R}^j \to \mathbb{R}^{d_u}$ satisfying (7.6).*

The rest of this section is dedicated to giving an idea of the proof of this crucial result. Note that, in the case where the drift system is strongly differentially observable of order d_x, i.e., in the context of Theorem 7.2, $\mathbf{H}_i$ is a submersion and thus opens for all $i \leq d_x$, so that the result holds.

A first important thing to notice is that the following property must be satisfied for the identity (7.6) to be satisfied (on $\mathcal{S}$).

Definition 7.1 *(Property $\mathcal{A}(i)$)* For a nonzero integer i, we will say Property $\mathcal{A}(i)$ holds if

$$L_g L_f^{i-1} h(x_a) = L_g L_f^{i-1} h(x_b) \qquad \forall (x_a, x_b) \in \mathcal{S}^2 \ : \ \mathbf{H}_i(x_a) = \mathbf{H}_i(x_b) \ .$$

Actually, the converse is true and is a direct consequence of Lemma A.10.

Lemma 7.1 *If Property $\mathcal{A}(i)$ is satisfied with $\mathcal{S}$ containing the given compact set $\mathcal{C}$, then there exists a continuous function $\mathfrak{g}_i : \mathbb{R}^i \to \mathbb{R}^{d_u}$ satisfying (7.6).*

Property $\mathcal{A}(i)$ being sufficient to obtain the existence of a function $\mathfrak{g}_i$ satisfying (7.6), we study now under which conditions it holds. Clearly, $\mathcal{A}(i)$ is satisfied for all $i \geq m$ if $\mathbf{H}_m$ is injective. If we do not have this injectivity property, the situation is more complex. To overcome the difficulty, we introduce the following property.

Definition 7.2 (*Property $\mathcal{B}(i)$*) For an integer $2 \leq i \leq d_x + 1$, we will say that Property $\mathcal{B}(i)$ holds if for any (x_a, x_b) in $\mathcal{S}^2$ such that $x_a \neq x_b$ and $\mathbf{H}_i(x_a) = \mathbf{H}_i(x_b)$, there exists a sequence $(x_{a,k}, x_{b,k})$ of points in $\mathcal{S}^2$ converging to (x_a, x_b) such that for all k, $\mathbf{H}_i(x_{a,k}) = \mathbf{H}_i(x_{b,k})$ and $\dfrac{\partial \mathbf{H}_{i-1}}{\partial x}$ is full-rank at $x_{a,k}$ or $x_{b,k}$.

As in this property, let $x_a \neq x_b$ be such that $\mathbf{H}_i(x_a) = \mathbf{H}_i(x_b)$. If $\frac{\partial \mathbf{H}_{i-1}}{\partial x}$ is full-rank at either x_a or x_b, then we can take $(x_{a,k}, x_{b,k})$ constant equal to (x_a, x_b). Thus, it is sufficient to check $\mathcal{B}(i)$ around points where neither $\frac{\partial \mathbf{H}_{i-1}}{\partial x}(x_a)$ nor $\frac{\partial \mathbf{H}_{i-1}}{\partial x}(x_b)$ is full-rank. But according to [5, Theorem 4.1], the set of points where $\frac{\partial \mathbf{H}_{d_x}}{\partial x}$ is not full-rank is of codimension at least one for a uniformly observable system. Thus, it is always possible to find points $x_{a,k}$ as close to x_a as we want such that $\frac{\partial \mathbf{H}_{i-1}}{\partial x}(x_{a,k})$ is full-rank. The difficulty of $\mathcal{B}(i)$ thus rather lies in ensuring that we have also $\mathbf{H}_i(x_{a,k}) = \mathbf{H}_i(x_{b,k})$. In Sect. 7.2.1.2, we prove:

Lemma 7.2 ([1]) *Suppose System (5.1) is uniformly instantaneously observable on $\mathcal{S}$. Then,*

- *Property $\mathcal{A}(1)$ is satisfied.*
- *If, for some i in $\{2, \ldots, d_x + 1\}$, Property $\mathcal{B}(i)$ holds and Property $\mathcal{A}(j)$ is satisfied for all j in $\{1, \ldots, i-1\}$, then Property $\mathcal{A}(i)$ holds.*

Thus, the first point in Theorem 7.4 is proved. Besides, a direct consequence of Lemmas 7.1 and 7.2 is that a sufficient condition to have the existence of the functions $\mathfrak{g}_i$ for i in $\{2, \ldots, d_x + 1\}$ is to have $\mathcal{B}(j)$ for j in $\{2, \ldots, i\}$. Actually, in [1], it is proved that $\mathscr{B}(j)$ is satisfied when $\mathbf{H}_j$ is an open map, which finishes the proof of Theorem 7.4. However, the following example shows that the openness of $\mathbf{H}_j$ is not necessary.

Example 7.2 Consider the system defined as

$$\begin{cases} \dot{x}_1 = x_2 \\ \dot{x}_2 = x_3^3 x_1 \\ \dot{x}_3 = 1 + u \end{cases} , \quad y = x_1 \tag{7.7}$$

On $\mathscr{S} = \left\{x \in \mathbb{R}^3 : x_1^2 + x_2^2 \neq 0\right\}$, and whatever u is, the knowledge of y and of its three first derivatives

$$\dot{y} = x_2 , \quad \ddot{y} = x_3^3 x_1 , \quad \dddot{y} = 3x_3^2 x_1 (1 + u) + x_3^3 x_2$$

gives us x_1, x_2, and x_3. Thus, the system is uniformly instantaneously observable on $\mathscr{S}$. Besides, the function

$$\mathbf{H}_4(x) = \begin{pmatrix} x_1 \\ x_2 \\ x_3^3 x_1 \\ 3x_3^2 x_1 + x_3^3 x_2 \end{pmatrix}$$

is injective on $\mathscr{S}$; thus, the drift system is weakly differentially observable of order 4 on $\mathscr{S}$. Now, although $\mathbf{H}_2$ is trivially an open map on $\mathscr{S}$, $\mathbf{H}_3$ is not. Indeed, consider for instance the open ball[11] $B_{\frac{1}{2}}(0, x_2, 0)$ in $\mathbb{R}^3$ for some x_2 such that $|x_2| > \frac{1}{2}$. $B_{\frac{1}{2}}(0, x_2, 0)$ is contained in $\mathscr{S}$. Suppose its image by $\mathbf{H}_3$ is an open set of $\mathbb{R}^3$. It contains $\mathbf{H}_3(0, x_2, 0) = (0, x_2, 0)$ and thus $(\varepsilon, x_2, \varepsilon)$ for any sufficiently small ε. This means that there exist x in $B_{\frac{1}{2}}(0, x_2, 0)$ such that $(\varepsilon, x_2, \varepsilon) = \mathbf{H}_3(x)$, i.e., necessarily $x_1 = \varepsilon$ and $x_3 = 1$. But this point is not in $B_{\frac{1}{2}}(0, x_2, 0)$, and we have a contradiction. Therefore, $\mathbf{H}_3$ is not open. However, $\mathscr{B}(3)$ trivially holds because $\mathbf{H}_2$ is full-rank everywhere. ▲

7.2.1.2 Proof of Lemma 7.2

Lemma 7.2 is fundamental for Theorem 7.4 and its important corollary Theorem 7.2. That is why we dedicate a whole section to its proof which was given in [1]. It is built in the same spirit as the proof of Theorem 7.2 in [4] but in a more detailed and complete way so that the reader can understand how the fact that $\mathbf{H}_{d_\xi}$ is no longer a diffeomorphism makes a great difference when going from Theorem 7.2 to 7.4.

Assume the system is uniformly instantaneously observable on $\mathscr{S}$. We first show that property $\mathscr{A}(1)$ holds. Suppose there exists (x_a^*, x_b^*) in $\mathscr{S}^2$ and k in $\{1, \ldots, d_u\}$ such that $x_a^* \neq x_b^*$ and

[11] $B_r(x)$ denotes the open ball centered at x and with radius r.

$$h(x_a^*) = h(x_b^*) \quad , \qquad L_{g_k} h(x_a^*) \neq L_{g_k} h(x_b^*) \; .$$

Then, the control law u with all its components zero except its kth one which is

$$u_k \;=\; -\frac{L_f h(x_a) - L_f h(x_b)}{L_{g_k} h(x_a) - L_{g_k} h(x_b)} \; .$$

is defined on a neighborhood of (x_a^*, x_b^*). The corresponding solutions $X(x_a^*; t; u)$ and $X(x_b^*; t; u)$ are defined on some time interval $[0, \bar{t})$ and satisfy

$$h(X(x_a^*; t; u)) = h(X(x_b^*; t; u)) \qquad \forall t \in [0, \bar{t}) \; .$$

Since x_a^* is different from x_b^*, this contradicts the instantaneous observability. Thus, $\mathscr{A}(1)$ holds.

Let now i in $\{2, \ldots, d_x + 1\}$ be such that Property $\mathscr{B}(i)$ holds and $\mathscr{A}(j)$ is satisfied for all j in $\{1, \ldots, i-1\}$. To establish by contradiction that $\mathscr{A}(i)$ holds, we assume this is not the case. This means that there exists $(x_{a,0}^*, x_{b,0}^*)$ in $\mathscr{S}^2$ and k in $\{1, \ldots, d_u\}$ such that $\mathbf{H}_i(x_{a,0}^*) = \mathbf{H}_i(x_{b,0}^*)$ but $L_{g_k} L_f^{i-1}(x_{a,0}^*) \neq L_{g_k} L_f^{i-1}(x_{b,0}^*)$. This implies $x_{a,0}^* \neq x_{b,0}^*$. By continuity of $L_{g_k} L_f^{i-1}$ and according to $\mathscr{B}(i)$, there exists x_a^* (resp x_b^*) in $\mathscr{S}$ sufficiently close to $x_{a,0}^*$ (resp $x_{b,0}^*$) satisfying $x_a^* \neq x_b^*$,

$$\mathbf{H}_i(x_a^*) = \mathbf{H}_i(x_b^*) \; , \quad L_{g_k} L_f^{i-1}(x_a^*) \neq L_{g_k} L_f^{i-1}(x_b^*) \; ,$$

and $\frac{\partial \mathbf{H}_{i-1}}{\partial x}$ is full-rank at x_a^* or x_b^*. Without loss of generality, we suppose it is full-rank at x_a^*. Thus, $\frac{\partial \mathbf{H}_j}{\partial x}$ is full-rank at x_a^* for all $j < i \leq d_x + 1$. We deduce that there exists an open neighborhood $\mathscr{V}_a$ of x_a^* such that for all $j < i$, $\frac{\partial \mathbf{H}_j}{\partial x}$ is full-rank on $\mathscr{V}_a$. Since $\mathscr{A}(j)$ holds for all $j < i$, according to Lemma A.11, $\mathbf{H}_j(\mathscr{V}_a)$ is open for all $j < i$ and there exist locally Lipschitz functions $\mathfrak{g}_j : \mathbf{H}_j(\mathscr{V}_a) \to \mathbb{R}^{d_u}$ such that, for all x_α in $\mathscr{V}_a$,

$$\mathfrak{g}_j(\mathbf{H}_j(x_\alpha)) = L_g L_f^{j-1} h(x_\alpha) \; . \tag{7.8}$$

Also, $\mathbf{H}_j(x_a^*) = \mathbf{H}_j(x_b^*)$ implies that $\mathbf{H}_j(x_b^*)$ is in the open set $\mathbf{H}_j(\mathscr{V}_a)$. Continuity of each $\mathbf{H}_j$ implies the existence of an open neighborhood $\mathscr{V}_b$ of x_b^* such that $\mathbf{H}_j(\mathscr{V}_b)$ is contained in $\mathbf{H}_j(\mathscr{V}_a)$ for all $j < i$. Thus, for any x_β in $\mathscr{V}_b$, $\mathbf{H}_j(x_\beta)$ is in $\mathbf{H}_j(\mathscr{V}_a)$, and there exists x_α in $\mathscr{V}_a$ such that $\mathbf{H}_j(x_\alpha) = \mathbf{H}_j(x_\beta)$. According to $\mathscr{A}(j)$, this implies that $L_g L_f^{j-1} h(x_\beta) = L_g L_f^{j-1} h(x_\alpha)$ and with (7.8),

$$L_g L_f^{j-1} h(x_\beta) = L_g L_f^{j-1} h(x_\alpha) = \mathfrak{g}_j(\mathbf{H}_j(x_\alpha)) = \mathfrak{g}_j(\mathbf{H}_j(x_\beta)) \; .$$

Therefore, (7.8) holds on $\mathscr{V}_a$ and $\mathscr{V}_b$.

Then, the control law u with all its components zero except its kth one which is

$$u_k \;=\; -\frac{L_f^i h(x_a) - L_f^i h(x_b)}{L_{g_k} L_f^{i-1} h(x_a) - L_{g_k} L_f^{i-1} h(x_b)}$$

is defined on a neighborhood of (x_a^*, x_b^*). The corresponding solutions $X(x_a^*; t; u)$ and $X(x_b^*; t; u)$ are defined on some time interval $[0, \bar{t})$ where they remain in $\mathscr{V}_a$ and $\mathscr{V}_b$, respectively. Let $Z_a(t) = \mathbf{H}_i(X(x_a^*; t; u))$, $Z_b(t) = \mathbf{H}_i(X(x_b^*; t; u))$, and $W(t) = Z_a(t) - Z_b(t)$ on $[0, \bar{t})$. Since, for all $j < i$, (7.8) holds on $\mathscr{V}_a$ and $\mathscr{V}_b$, (W, Z_a) is solution to the system:

$$\begin{cases} \dot{w}_1 &= w_2 + (\mathfrak{g}_1(\xi_{a,1}) - \mathfrak{g}_1(\xi_{a,1} - w_1))\, u \\ &\vdots \\ \dot{w}_j &= w_{j+1} + (\mathfrak{g}_j(\xi_{a,1}, \dots, \xi_{a,j}) - \mathfrak{g}_j(\xi_{a,1} - w_1, \dots, \xi_{a,j} - w_j))\, u \\ &\vdots \\ \dot{w}_i &= 0 \\ \dot{\xi}_{a,1} &= \xi_2 + \mathfrak{g}_1(\xi_{a,1})\, u \\ &\vdots \\ \dot{\xi}_{a,j} &= \xi_{j+1} + \mathfrak{g}_j(\xi_{a,1}, \dots, \xi_{a,j})\, u \\ &\vdots \\ \dot{\xi}_{a,i} &= \tilde{u} \end{cases}$$

with initial condition $(0, \mathbf{H}_i(x_a^*))$, where $\tilde{u}$ is the time derivative of $Z_{a,i}(t)$. Note that the function $(0, Z_a)$ is also a solution to this system with the same initial condition. Since the functions involved in this system are locally Lipschitz, it admits a unique solution. Hence, for all t in $[0, \bar{t}[$, $W(t) = 0$, and thus $Z_a(t) = Z_b(t)$, which implies $h(X(x_a^*, t)) = h(X(x_b^*, t))$. Since x_a^* is different from x_b^*, this contradicts the uniform observability. Thus, $\mathscr{A}(i)$ holds.

The key part of this proof is to ensure that (7.8) holds both around x_a^* and x_b^* with the same $\mathfrak{g}_j$, so that $(0, Z_a)$ is solution to the system above. For that, we have used the openness of $\mathbf{H}_j$ around x_a^*, which is automatically satisfied when $\mathbf{H}_{d_\xi}$ is a diffeomorphism.

7.2.1.3 A Solution to Problem 7.1

With Theorems 7.3 and 7.4, we have the following solution to Problem 7.1.

Theorem 7.5 ([1]) *Let $\mathscr{S}$ be an open set containing the given compact set $\mathscr{C}$. Suppose*

- *System (5.1) is uniformly instantaneously observable on $\mathscr{S}$.*
- *The drift system of System (5.1) is weakly differentially observable of order m on $\mathscr{S}$.*

With selecting $T = \mathbf{H}_\mathbf{m}$ and $d_\xi = m$, we have a solution to Problem 7.1 if we pick either $\daleth = 1$, or $\daleth = i$ when $\mathbf{H}_j$ is an open map for any j in $\{2, \dots, i\}$ with $i \leq d_x$.

Remark 7.2

- As seen in Example 7.2, the openness of the functions $\mathbf{H}_j$ is sufficient but not necessary. We may ask only for $\mathscr{B}(j)$ for any j in $\{2, \dots, i\}$ with $i \leq d_x + 1$. Besides, this weaker assumption allows to obtain the existence of $\mathfrak{g}_i$ up to the order $d_x + 1$.
- Consider the case where $\mathscr{B}(j)$ is satisfied for all $j \leq d_x + 1$ and $m = d_x + 2$. Then, we have $\mathtt{T} = d_x + 1$ and it is possible to obtain a full triangular form of dimension $d_\xi = \mathtt{T} + 1 = m = d_x + 2$. Actually, we still have a full triangular form if we choose $d_\xi > m$. Indeed, $\mathbf{H}_m$ being injective, $\mathscr{A}(i)$ is satisfied for all i larger than m; thus, there also exist continuous functions $\mathfrak{g}_i : \mathbb{R}^i \to \mathbb{R}^{d_u}$ satisfying (7.6) for all $i \geq m$. It follows that $\mathtt{T}$ can be taken larger than $d_x + 1$ and $d_\xi = \mathtt{T} + 1$ larger than m.
- If Problem 7.1 is solved with $d_\xi = \mathtt{T} + 1$, we have a full triangular normal form of dimension d_ξ. But, at this point we know nothing about the regularity of the functions $\mathfrak{g}_i$, besides continuity. As we saw in Example 7.1, even the usual assumption of strong differential observability is not sufficient to make it Lipschitz everywhere. As studied in Chap. 4, this may impede the convergence of a high-gain observer. That is why, in the next section, we look for conditions under which the Lipschitzness is ensured.
- As explained in Sect. 7.1.1, another way of solving Problem 7.1 is to allow the transformation T to depend on the control u and its derivatives. In particular, if $d_\xi > \mathtt{T} + 1$, a full triangular form may still be obtained with $T = (\mathbf{H}_\mathtt{T}, \tilde{T})$ where the components $\tilde{T}_i$ of $\tilde{T}$ are defined recursively as

$$\tilde{T}_1 = L_f^{\mathtt{T}} h \quad , \quad \tilde{T}_{i+1} = L_{f+gu} \tilde{T}_i + \sum_{j=0}^{i-2} \frac{\partial \tilde{T}_i}{\partial u^{(j)}} u^{(j+1)}$$

until (if possible) the map $x \mapsto T(x, u, \dot{u}, \dots)$ becomes injective for all $(u, \dot{u}, \dots)$. The interest of Theorem 7.5 is to ensure triangularity while reducing the order of differentiation of u compared to Theorem 7.1.

Example 7.3 Coming back to Example 7.2, we have seen that $\mathbf{H}_2$ is open and that $\mathbf{H}_3$ is not but $\mathscr{B}(3)$ is satisfied. Besides, the system is weakly differentially observable of order 4. We deduce that there exists a full triangular form of order 4. Indeed, we have $L_g h(x) = L_g L_f h(x) = 0$ and

$$L_g L_f^2 h(x) = 3x_3^2 x_1 = 3(L_f^2 h(x))^{\frac{2}{3}} (h(x))^{\frac{1}{3}}$$

so that we can take

$$\mathfrak{g}_1 = \mathfrak{g}_2 = 0 \quad , \quad \mathfrak{g}_3(\xi_1, \xi_2, \xi_3) = 3\xi_3^{\frac{2}{3}} \xi_1^{\frac{1}{3}} .$$

As for φ_4 and $\mathfrak{g}_4$, they are obtained via inversion of $\mathbf{H}_4$, i.e., on $\mathbb{R}^4 \setminus \{(0, 0, \xi_3), \ \xi_3 \in \mathbb{R}\}$

$$\mathbf{H}_4^{-1}(\xi) = \left(\xi_1, \xi_2, \left(\frac{(\xi_4 - 3\xi_3^{\frac{2}{3}}\xi_1^{\frac{1}{3}})^2 + \xi_3^2}{\xi_1^2 + \xi_2^2}\right)^{\frac{1}{6}}\right).$$

▲

7.2.2 *Lipschitzness of the Triangular Form*

7.2.2.1 A Sufficient Condition

We saw with Examples 7.1 and 7.2 that uniform instantaneous observability is not sufficient for the functions $\mathfrak{g}_i$ to be Lipschitz. Nevertheless, we are going to show in this section that it is sufficient except maybe around the image of points where $\frac{\partial \mathbf{H}_i}{\partial x}$ is not full-rank ($x_1 = 0$ or $x_3 = 0$ in Example 7.2).

Consider the open set $\mathscr{R}_i$ of points in $\mathscr{S}$ where $\frac{\partial \mathbf{H}_i}{\partial x}$ has full-rank. According to [9, Corollaire pp. 68–69], if $\mathbf{H}_i$ is an open map, $\mathscr{R}_i$ is an open dense set. Anyway, assume $\mathscr{R}_{d_x} \cap \mathscr{C}$ is nonempty. Then, there exists $\varepsilon_0 > 0$ such that, for all ε in $(0, \varepsilon_0]$, the set

$$K_{i,\varepsilon} = \left\{x \in \mathscr{R}_i \cap \mathscr{C}\ ,\quad d(x\ ,\ \mathbb{R}^{d_x} \backslash \mathscr{R}_i) \geq \varepsilon\right\}\ .$$

is nonempty and compact, and such that its points are (ε)-away from singular points. The next theorem shows that the functions $\mathfrak{g}_i$ can be taken Lipschitz on the image of $K_{i,\varepsilon}$, i.e., everywhere except arbitrary close to the image of points where the rank of the Jacobian of $\mathbf{H}_i$ drops.

Theorem 7.6 ([1]) *Assume System (5.1) is uniformly instantaneously observable on an open set $\mathscr{S}$ containing the compact set $\mathscr{C}$. For all i in $\{1, \ldots, d_x\}$ and for any ε in $(0, \varepsilon_0]$, there exists a Lipschitz function $\mathfrak{g}_i : \mathbb{R}^i \to \mathbb{R}^{d_u}$ satisfying (7.6) for all x in $K_{i,\varepsilon}$.*

Proof As noticed after the statement of Property $\mathscr{B}(i)$, since $\frac{\partial \mathbf{H}_i}{\partial x}$ has full-rank in the open set $\mathscr{R}_i$, Property $\mathscr{B}(i)$ holds on $\mathscr{R}_i$ (i.e., with $\mathscr{R}_i$ replacing $\mathscr{S}$ in its statement). It follows from Lemma 7.2 that $\mathscr{A}(i)$ is satisfied on $\mathscr{R}_i$. Besides, according to Lemma A.11, $\mathbf{H}_i(\mathscr{R}_i)$ is open and there exists a C^1 function $\mathfrak{g}_i$ defined on $\mathbf{H}_i(\mathscr{R}_i)$ such that for all x in $\mathscr{R}_i$, $\mathfrak{g}_i(\mathbf{H}_i(x)) = L_g L_f^{i-1} h(x)$. Now, $K_{i,\varepsilon}$ being a compact set contained in $\mathscr{R}_i$, and $\mathbf{H}_i$ being continuous, $\mathbf{H}_i(K_{i,\varepsilon})$ is a compact set contained in $\mathbf{H}_i(\mathscr{R}_i)$. Thus, $\mathfrak{g}_i$ is Lipschitz on $\mathbf{H}_i(K_{i,\varepsilon})$. According to [10], there exists a Lipschitz extension of $\mathfrak{g}_i$ to $\mathbb{R}^i$ coinciding with $\mathfrak{g}_i$ on $\mathbf{H}_i(K_{i,\varepsilon})$ and thus verifying (7.6) for all x in $K_{i,\varepsilon}$. □

If the drift system is strongly differentially observable of order $m = d_x$ on $\mathscr{S}$, the Jacobian of $\mathbf{H}_i$ for any i in $\{1, \ldots, d_x\}$ has full-rank on $\mathscr{S}$. Thus, taking $d_\xi =$

$\mathtt{T}+1=m=d_x$ a full Lipschitz triangular form of dimension d_x exists; i.e., we recover the result of Theorem 7.2.

Example 7.4 In Example 7.2, $\mathbf{H}_3$ is full-rank on $\mathscr{S} \setminus \{x \in \mathbb{R}^3 \mid x_1 = 0 \text{ or } x_3 = 0\}$. Thus, according to Theorem 7.6, the only points where $\mathfrak{g}_3$ may not be Lipschitz are the image of points where $x_1 = 0$ or $x_3 = 0$. Let us study more precisely what happens around those points. Take $x_a = (x_{1,a}, x_{2,a}, 0)$ in $\mathscr{S}$. If there existed a locally Lipschitz function $\mathfrak{g}_3$ verifying (7.6) around x_a, there would exist $\mathfrak{a} > 0$ such that for any $x_b = (x_{1,b}, x_{2,a}, x_{3,b})$ sufficiently close to x_a with $x_{1,b} \neq 0$, $|3x_{3,b}^2| \leq \mathfrak{a}|x_{3,b}^3|$, which we know is impossible. Therefore, there does not exist a function $\mathfrak{g}_3$ which is Lipschitz around the image of points where $x_3 = 0$. Let us now study what happens elsewhere, namely on $\tilde{\mathscr{S}} = \mathscr{S} \setminus \{x \in \mathbb{R}^3 \,|x_3 = 0\}$. It turns out that, on any compact set $\mathscr{C}$ of $\tilde{\mathscr{S}}$, there exists[12] $\mathfrak{a}$ such that we have for all (x_a, x_b) in $\mathscr{C}^2$,

$$|x_{3,a}^2 x_{1,a} - x_{3,b}^2 x_{1,b}| \leq \mathfrak{a}(|x_{1,a} - x_{1,b}| + |x_{3,a}^3 x_{1,a} - x_{3,b}^3 x_{1,b}|)$$

Therefore, the continuous function $\mathfrak{g}_3$ found earlier in Example 7.3 such that $\mathfrak{g}_3(\mathbf{H}_3(x)) = L_g L_f^2(x) = 3x_3^2 x_1$ on $\mathscr{S}$ (and thus on $\mathscr{C}$) verifies in fact

$$|\mathfrak{g}_3(\xi_a) - \mathfrak{g}_3(\xi_b)| \leq \mathfrak{a}|\xi_a - \xi_b|$$

on $\mathbf{H}_3(\mathscr{C})$ and can be extended to a Lipschitz function on $\mathbb{R}^3$ according to [10]. We conclude that although $\mathbf{H}_3$ does not have a full-rank Jacobian everywhere on $\mathscr{C}$ (singularities at $x_1 = 0$), it is possible to find a Lipschitz function $\mathfrak{g}_3$ solution to our problem on this set. ▲

7.2.2.2 A Necessary Condition

We have just seen that the condition in Theorem 7.6 that the Jacobian of $\mathbf{H}_i$ is full-rank is sufficient but not necessary. In order to have locally Lipschitz functions $\mathfrak{g}_i$ satisfying (7.6), there must exist for all x a strictly positive number $\mathfrak{a}$ such that for all (x_a, x_b) in a neighborhood of x,

$$|L_g L_f^{i-1} h(x_a) - L_g L_f^{i-1} h(x_b)| \leq \mathfrak{a}\,|\mathbf{H}_i(x_a) - \mathbf{H}_i(x_b)|\,. \tag{7.9}$$

We have the following necessary condition:

---

[12] If $x_{1,a}$ and $x_{1,b}$ are both zero, the inequality is trivial. Suppose $|x_{1,a}| > |x_{1,b}|$, and denote $\rho = \frac{x_{1,b}}{x_{1,a}}$. If $\rho < 0$, we have directly $|x_{3,a}^2 - \rho\, x_{3,b}^2| \leq \max\{x_{3,a}^2, x_{3,b}^2\}|1-\rho|$. If now $\rho > 0$, $x_{3,a}^2 - \rho\, x_{3,b}^2 = \frac{(x_{3,a}^3 - \rho^{\frac{3}{2}}\, x_{3,b}^3)(x_{3,a} + \sqrt{\rho} x_{3,b})}{x_{3,a}^2 + \sqrt{\rho} x_{3,a} x_{3,b} + \rho x_{3,b}^2}$ and thus $|x_{3,a}^2 - \rho\, x_{3,b}^2| \leq \frac{2\sqrt{2}}{\sqrt{x_{3,a}^2 + \rho x_{3,a}^2}} |x_{3,a}^3 - \rho^{\frac{2}{3}}\, x_{3,b}^3|$. Besides, $|x_{3,a}^3 - \rho^{\frac{3}{2}}\, x_{3,b}^3| = |x_{3,a}^3 - \rho x_{3,b}^3 + \rho(1-\sqrt{\rho})\, x_{3,b}^3| \leq |x_{3,a}^3 - \rho x_{3,b}^3| + \frac{\rho |x_{3,b}^3|}{1+\sqrt{\rho}} |1-\rho|$ which gives $\mathfrak{a}$ on compact sets.

Lemma 7.3 *Consider x in $\mathcal{S}$ such that* (7.9) *is satisfied in a neighborhood of x. Then, for any nonzero vector v in $\mathbb{R}^{d_x}$, and any k in $\{1, \ldots, d_u\}$, we have:*

$$\frac{\partial \mathbf{H}_i}{\partial x}(x)\, v = 0 \quad \Rightarrow \quad \frac{\partial L_{g_k} L_f^{i-1} h}{\partial x}(x)\, v = 0\,. \tag{7.10}$$

Proof Assume there exists a nonzero vector v in $\mathbb{R}^{d_x}$ such that $\dfrac{\partial \mathbf{H}_i}{\partial x}(x)\, v = 0$. Choose $r > 0$ such that Inequality (7.9) holds on $B_r(x)$, the ball centered at x and of radius r. Consider for any integer p the vector x_p in $B_r(x)$ defined by $x_p = x - \frac{1}{p}\frac{1}{|v|}v$. This gives a sequence converging to x when p tends to infinity. We have

$$0 \leq \frac{|L_{g_k} L_f^{i-1} h(x) - L_{g_k} L_f^{i-1} h(x_p)|}{|x - x_p|} \leq \mathfrak{a}\, \frac{|\mathbf{H}_i(x) - \mathbf{H}_i(x_p)|}{|x - x_p|} \tag{7.11}$$

But, $\frac{\mathbf{H}_i(x) - \mathbf{H}_i(x_p)}{|x - x_p|}$ tends to $\frac{\partial \mathbf{H}_i}{\partial x}(x)\, v$ which by assumption is 0. Similarly, $\frac{1}{|x-x_p|}(L_{g_k} L_f^{i-1} h(x) - L_{g_k} L_f^{i-1} h(x_p))$ tends to $\dfrac{\partial L_{g_k} L_f^{i-1} h}{\partial x}(x)\, v$ which is also 0 according to (7.11). □

We conclude that when $\mathbf{H}_i$ does not have a full-rank Jacobian, it must satisfy Condition (7.10) to allow the existence of locally Lipschitz triangular functions $\mathfrak{g}_i$. This condition is in fact about uniform infinitesimal observability.

Definition 7.3 See [5, Definition I.2.1.3]. Consider the system lifted to the tangent bundle ([5, page 10])

$$\begin{cases} \dot{x} = f(x) + g(x)u \\ \dot{v} = \left[\frac{\partial f}{\partial x}(x) + \frac{\partial gu}{\partial x}(x)\right] v \end{cases}, \quad \begin{cases} y = h(x) \\ w = \frac{\partial h}{\partial x}(x)v \end{cases} \tag{7.12}$$

with v in $\mathbb{R}^{d_x}$ and w in $\mathbb{R}$ and the solutions of which are denoted $(X(x; t; u), V((x, v); t; u))$. System (5.1) is *uniformly instantaneously infinitesimally observable* on $\mathcal{S}$ if, for any pair (x, v) in $\mathcal{S} \times \mathbb{R}^{d_x} \setminus \{0\}$, any strictly positive number $\bar{t}$, and any C^1 function u defined on an interval $[0, \bar{t})$, there exists a time $t < \bar{t}$ such that $\dfrac{\partial h}{\partial x}(X(x; t; u))\, V((x, v); t; u) \neq 0$ and such that $X(x; s; u) \in \mathcal{S}$ for all $s \leq t$.

We have the following result.

Theorem 7.7 ([1]) *Suppose that the drift system of System (5.1) is strongly differentially observable of order m (or at least that $\mathbf{H}_m$ is an immersion on $\mathcal{S}$) and that Inequality* (7.9) *is verified at least locally around any point x in $\mathcal{S}$ for any i in $\{1, \ldots, m\}$. Then, the System (5.1) is uniformly infinitesimally observable on $\mathcal{S}$.*

Proof According to Lemma 7.3, we have (7.10). Now, take x in $\mathcal{S}$ and a nonzero vector v and suppose that there exists $\bar{t} > 0$ such that for all t in $[0, \bar{t})$, $X(x; t; u)$ is in $\mathcal{S}$ and $w(t) = \frac{\partial h}{\partial x}(X(x;t;u))V((x,v);t;u) = 0$. To simplify the notations, we denote $X(t) = X(x;t;u)$ and $V(t) = V((x,v);t;u)$. For all integer i, we denote

$$w_i(t) = \frac{\partial L_f^{i-1}h}{\partial x}(X(t))V(t) .$$

We note that for any function $\psi : \mathbb{R}^n \to \mathbb{R}$, we have

$$\overbrace{\frac{\partial \psi}{\partial x}(X(t))V(t)}^{\cdot} = \frac{\partial L_f \psi}{\partial x}(X(t))V(t) + \sum_{k=1}^{d_u} u_k \frac{\partial L_{g_k}\psi}{\partial x}(X(t))V(t) .$$

We deduce for all integer i and all t in $[0,\overline{t})$

$$\dot{w}_i(t) = w_{i+1}(t) + \sum_{k=1}^{d_u} u_k \frac{\partial L_{g_k} L_f^{i-1}h}{\partial x}(X(t))V(t) .$$

Let us show by induction that $w_i(t) = 0$ for all integer i and all t in $[0,\overline{t})$. It is true for $i = 1$ by assumption. Now, take an integer $i > 1$, and suppose $w_j(t) = 0$ for all t in $[0,\overline{t})$ and all $j \leq i$, i.e., $\frac{\partial \mathbf{H}_i}{\partial x}(X(x;t;u))V((x,v);t;u) = 0$ for all $t < \overline{t}$. In particular, $\dot{w}_i(t) = 0$ for all $t < \overline{t}$. Besides, according to (7.10), $\frac{\partial L_{g_k} L_f^{i-1}h}{\partial x}(X(x;t;u))V((x,v);t;u) = 0$ for all k in $\{1,\ldots,d_u\}$ and for all $t < \overline{t}$. Thus, $w_{i+1}(t) = 0$ for all $t < \overline{t}$. We conclude that w_i is zero on $[0,\overline{t}[$ for all i and in particular at time 0, $\frac{\partial \mathbf{H}_m}{\partial x}(x)v = (w_1(0),\ldots,w_m(0)) = 0$. But $\mathbf{H}_m$ is an immersion on $\mathcal{S}$; thus, necessarily $v = 0$ and we have a contradiction. □

Example 7.5 We go on with Example 7.2. The linearization of the dynamics (7.7) yields

$$\begin{cases} \dot{v}_1 = v_2 \\ \dot{v}_2 = x_3^3 v_1 + 3x_3^2 x_1 v_3 \\ \dot{v}_3 = 0 \end{cases} , \quad w = v_1 \tag{7.13}$$

Consider $x_0 = (x_1, x_2, 0)$ in $\mathcal{S}$ and $v_0 = (0, 0, v_3)$ with v_3 a nonzero real number. The solution to (7.7)–(7.13) initialized at (x_0, v_0) and with a constant input $u = -1$ is such that $X(x_0;t;u)$ remains in $\mathcal{S}$ in $[0,\overline{t})$ for some strictly positive $\overline{t}$ and $w(t) = 0$ for all t in $[0,\overline{t})$. Since v_0 is nonzero, System (7.7) is not uniformly instantaneously infinitesimally observable on $\mathcal{S}$. But, for System (7.7), $\mathbf{H}_7$ is an immersion on $\mathcal{S}$. We deduce from Theorem 7.7 that Inequality (7.9) is not satisfied for all i; i.e., there does not exist Lipschitz triangular functions $\mathfrak{g}_i$ for all i on $\mathcal{S}$. This is consistent with the conclusion of Example 7.4. However, on $\tilde{\mathcal{S}}$, i.e., when we remove the points where $x_3 = 0$, the system becomes uniformly instantaneously infinitesimally observable. Indeed, it can easily be checked that for x in $\tilde{\mathcal{S}}$, $w = \dot{w} = \ddot{w} = w^{(3)} = 0$ implies necessarily $v = 0$. Unfortunately, from our results, we cannot infer from this that the functions $\mathfrak{g}_i$ can be taken Lipschitz on $\tilde{\mathcal{S}}$. Nevertheless, the conclusion of Example 7.4 is that $\mathfrak{g}_3$ can be taken Lipschitz even around points with $x_1 = 0$. All this suggests a possible tighter link between uniform instantaneous infinitesimal observability and Lipschitzness of the triangular form. ▲

We conclude from this section that uniform instantaneous infinitesimal observability is required to have the Lipschitzness of the functions $\mathfrak{g}_i$ when they exist. However, we do not know if it is sufficient yet.

7.2.3 Back to Example 4.1

Consider the system

$$\begin{cases} \dot{x}_1 = x_2 \\ \dot{x}_2 = -x_1 + x_3^5 x_1 \\ \dot{x}_3 = -x_1 x_2 + u \end{cases} , \quad y = x_1 . \tag{7.14}$$

It would lead us too far from the main subject of this book to study here the solutions' behavior of this system. We note, however, that when u is zero, they evolve in the two-dimensional surface $\{x \in \mathbb{R}^3 : 3x_1^2 + 3x_2^2 + x_3^6 = c^6\}$. The equilibrium $(0, 0, x_3)$ being unstable at least for $c > 1$, we can hope for the existence of solutions remaining in the compact set

$$\mathscr{C}_{r,\varepsilon} = \left\{ x \in \mathbb{R}^3 : x_1^2 + x_2^2 \geq \varepsilon , \ 3x_1^2 + 3x_2^2 + x_3^6 \leq r \right\}$$

for instance when u is a small periodic time function, except maybe for pairs of input u and initial condition (x_1, x_2, x_3) for which resonance could occur.

On $\mathscr{S} = \left\{ x \in \mathbb{R}^3 : x_1^2 + x_2^2 \neq 0 \right\}$, and whatever u is, the knowledge of the function $t \mapsto y(t) = X_1(x, t)$ and therefore of its three first derivatives

$$\begin{aligned} \dot{y} &= x_2 \\ \ddot{y} &= -x_1 + x_3^5 x_1 \\ \dddot{y} &= -x_2 - 5x_3^4 x_1^2 x_2 + x_3^5 x_2 + 5x_3^4 x_1 u \end{aligned}$$

gives us x_1, x_2, and x_3. Thus, System (7.14) is uniformly instantaneously observable on $\mathscr{S}$. Besides, the function

$$\mathbf{H}_4(x) = \begin{pmatrix} x_1 \\ x_2 \\ -x_1 + x_3^5 x_1 \\ -x_2 - 5x_3^4 x_1^2 x_2 + x_3^5 x_2 \end{pmatrix}$$

is injective on $\mathscr{S}$ and admits the following left inverse, defined on $\left\{ \xi \in \mathbb{R}^4 : \xi_1^2 + \xi_2^2 \neq 0 \right\}$:

$$\mathbf{H}_4^{-1}(\xi) = \begin{pmatrix} \xi_1 \\ \xi_2 \\ \left(\frac{(\xi_3+\xi_1)\xi_1+\left[(\xi_4+\xi_2)+3\left|(\xi_3+\xi_1)\lfloor\xi_1\rceil^{\frac{3}{2}}\right|^{\frac{4}{5}}\xi_2\right]\xi_2}{\xi_1^2+\xi_2^2}\right)^{\frac{1}{5}} \end{pmatrix}$$

However, $\mathbf{H}_4$ is not an immersion because of a singularity of its Jacobian at $x_3 = 0$. So, the drift system is weakly differentially observable of order 4 on $\mathscr{S}$ but not strongly. The reader may check that it can be transformed into the continuous triangular normal form of dimension 4 given by (4.13).

7.3 General Lipschitz Triangular Form

Consider a general multi-input single-output control-affine system

$$\dot{x} = f(x) + g(x)u \quad , \quad y = h(x) + h^u(x)u \in \mathbb{R} \tag{7.15}$$

where g and h^u are matrix fields with values in $\mathbb{R}^{d_x \times d_u}$ and $\mathbb{R}^{1\times d_u}$ such that for any $u = (u_1, \ldots, u_{d_u})$ in $\mathbb{R}^{d_u}$,

$$g(x)u = \sum_{k=1}^{d_u} g_k(x)u_k \quad , \quad h^u(x)u = \sum_{k=1}^{d_u} h_k^u(x)u_k$$

with g_k vector fields of $\mathbb{R}^{d_x}$ and h_k^u real-valued functions. We want to know under which conditions this system can be transformed into a general Lipschitz triangular form (4.14)

$$\begin{cases} \dot{\xi}_1 = A_1(u, y)\,\xi_2 + \Phi_1(u, \xi_1) \\ \quad\vdots \\ \dot{\xi}_i = A_i(u, y)\,\xi_{i+1} + \Phi_i(u, \xi_1, \ldots, \xi_i) \\ \quad\vdots \\ \dot{\xi}_m = \Phi_m(u, \xi) \end{cases} \quad , \quad y = C_1(u)\xi_1$$

for which a Kalman high-gain observer (4.18) may exist.[13] Before stating the main result, we need some definitions introduced in [6].

Definition 7.4 The *observation space* of System (7.15), denoted $\mathscr{O}$, is the smallest real vector space such that

- $x \mapsto h(x)$ and $x \mapsto h_k^u(x)$ for any k in $\{1, \ldots, d_u\}$ are in $\mathscr{O}$.

[13] An additional excitation condition on the input is needed; see Chap. 4.

- $\mathscr{O}$ is stable under the Lie derivative along the vector fields f, g_1, ..., g_{d_u}; i.e., for any element ϕ of $\mathscr{O}$, $L_f\phi$ and $L_{g_k}\phi$ for all k in $\{1, \ldots, d_u\}$ are in $\mathscr{O}$.

We denote $d\mathscr{O}$ the codistribution of $\mathbb{R}^{d_x}$ defined by

$$d\mathscr{O}(x) = \left\{d\phi(x)\,, \quad \phi \in \mathscr{O}\right\}.$$

This leads to the following observability notion.

Definition 7.5 System (7.15) is said to satisfy the *observability rank condition* at a point x in $\mathbb{R}^{d_x}$ (resp on $\mathscr{S}$) if

$$\dim(d\mathscr{O}(x)) = d_x \quad (\text{resp } \forall x \in \mathscr{S})\,.$$

It is proved in [6] that the observability rank condition is sufficient to ensure the so-called local weak observability, which roughly means that any point can be instantaneously distinguished from its neighbors via the output. In fact, this property is also necessary on a dense subset of $\mathscr{X}$. We refer the interested reader to [6] for a more precise account of those notions.

In [2], the authors relate the observability rank condition to the ability of transforming (at least locally) a system into a general Lipschitz triangular form.

Theorem 7.8 ([2]) *If System* (7.15) *satisfies the observability rank condition at* x_0, *then there exists a neighborhood* $\mathscr{V}$ *of* x_0 *and an injective immersion* T *on* $\mathscr{V}$ *which transforms System* (7.15) *into a general Lipschitz triangular form* (4.14) *on* $\mathscr{V}$ *with the linear parts* A_i *independent from the output, i.e.,* $A_i(u, y) = A_i(u)$.

This result is local because the rank condition is of local nature and does not say that we can select the same immersion T around every point of $\mathscr{X}$, let alone that this function is injective on $\mathscr{X}$. However, we give this result all the same because the idea of the construction of the function T is the same whether we look for a global immersion or a local one. Here is the algorithm presented in [2]:

1. Take $T^1(x) = (h(x), h_1^u(x), \ldots, h_{d_u}^u(x))$ of dimension $N_1 = d_u + 1$.
2. Suppose $T^1, \ldots, T^i$ have been constructed in the previous steps, of dimension $N_1, \ldots, N_i$. Pick among their $N_1 + \cdots + N_i$ differentials a maximum number ν_i of differentials $d\phi_1, \ldots, d\phi_{\nu_i}$ which generate a regular codistribution around x_0; i.e., there exists a neighborhood of x_0 where $\dim(\mathsf{span}\{d\phi_1(x), \ldots, d\phi_{\nu_i}(x)\})$ is constant and equal to ν_i.
   - If $\nu_i = d_x$ stop ;
   - Otherwise, build T^{i+1} with every functions $L_f T_j^i$ and $L_{g_k} T_j^i$, with j in $\{1, \ldots, N_i\}$ and k in $\{1, \ldots, d_u\}$, except those whose differential already belongs to $\mathsf{span}\{d\phi_1(x), \ldots, d\phi_{\nu_i}(x)\}$ in a neighborhood of x_0.

Finally, denoting m the number of iterations, take $T(x) = (T^1(x), \ldots, T^m(x))$.

The observability rank condition ensures that the algorithm stops at some point because computing the successive T^i comes back to progressively generating all $\mathcal{O}$ which is of dimension d_x around x_0. Besides, it is shown in [2] that when the differential $d\phi$ of some real-valued function ϕ is such that, in a neighborhood of x_0, $d\phi(x)$ belongs to $\mathsf{span}\{d\phi_1(x), \ldots, d\phi_{\nu_i}(x)\}$ with $d\phi_1(x), \ldots, d\phi_{\nu_i}(x)$ independent, then ϕ can be locally expressed in a Lipschitz way in terms of $\phi_1, \ldots, \phi_{\nu_i}$. Therefore, either the derivatives of the elements of T^i are in T^{i+1} or they can be expressed in terms of the previous $T^1, \ldots, T^i$. It follows that for any i, there exist a matrix $A_i(u)$ and a function Φ_i (linear in u and with $\Phi(u, \cdot)$ Lipschitz) such that

$$\dot{\overline{T^i(x)}} = L_f T^i(x) + \sum_{k=1}^{d_u} u_k L_{g_k} T^i(x) = A_i(u) T^{i+1}(x) + \Phi_i(u, T^1(x), \ldots, T^i(x)),$$

which gives the general triangular form (4.14).

Note that the transformation T thus obtained is a local immersion. If we are interested in a global transformation, the same algorithm can be applied but everything must be checked globally (and not in a neighborhood of x_0) and we need to go on with this algorithm until obtaining a global injective immersion. But there is no guarantee that this will be possible, unless a stronger assumption is made. In particular, if the drift system (i.e., with $u \equiv 0$) is strongly differentially observable of some order p, the algorithm provides a global injective immersion in a maximum of p iterations. Beware, however, that it still remains to check that the functions Φ_i exist globally. If this is not the case, it is always possible to put the corresponding $L_f T^i_j(x)$ or $L_{g_k} T^i_j(x)$ in T^{i+1}, but this is bound to considerably increase the dimension of T (and thus of the observer).

Finally, it is important to remark that this design enables to avoid the strong assumption of uniform observability needed for the classical triangular form, by stuffing the $L_{g_k} T^i_j(x)$ which do not verify the triangularity constraint into the state. The first obvious setback is that it often leads to observers of very large dimension. But mostly, unlike the classical Lipschitz triangular form which admits a high-gain observer without further assumption, the possibility of observer design for the general Lipschitz triangular form is not automatically achieved as seen in Chap. 4: Building the transformation is not enough, and one needs to check an additional excitation condition on the input.

References

1. Bernard, P., Praly, L., Andrieu, V., Hammouri, H.: On the triangular canonical form for uniformly observable controlled systems. Automatica **85**, 293–300 (2017)
2. Besançon, G., Ticlea, A.: An immersion-based observer design for rank-observable nonlinear systems. IEEE Trans. Autom. Control **52**(1), 83–88 (2007)
3. Gauthier, J.P., Bornard, G.: Observability for any u(t) of a class of nonlinear systems. IEEE Trans. Autom. Control **26**, 922–926 (1981)

4. Gauthier, J.P., Hammouri, H., Othman, S.: A simple observer for nonlinear systems application to bioreactors. IEEE Trans. Autom. Control **37**(6), 875–880 (1992)
5. Gauthier, J.P., Kupka, I.: Deterministic Observation Theory and Applications. Cambridge University Press, Cambridge (2001)
6. Hermann, R., Krener, A.: Nonlinear controllability and observability. IEEE Trans. Autom. Control **22**(5), 728–740 (1977)
7. Jouan, P., Gauthier, J.: Finite singularities of nonlinear systems. Output stabilization, observability, and observers. J. Dyn. Control Syst. **2**(2), 255–288 (1996)
8. Kirszbraun, M.D.: Über die zusammenziehende und Lipschitzsche transformationen. Fundam. Math. **22**, 77–108 (1934)
9. Leborgne, D.: Calcul Différentiel et Géometrie. Presse Universitaire de France, France (1982)
10. McShane, E.J.: Extension of range of functions. Bull. Am. Math. Soc. **40**(12), 837–842 (1934)
11. Rapaport, A., Maloum, A.: Design of exponential observers for nonlinear systems by embedding. Int. J. Robust Nonlinear Control **14**, 273–288 (2004)
12. Valentine, F.A.: A Lipschitz condition preserving extension for a vector function. Am. J. Math. **67**(1), 83–93 (1945)
13. Zeitz, M.: Observability canonical (phase-variable) form for nonlinear time-variable systems. Int. J. Syst. Sci. **15**(9), 949–958 (1984)

Part III
Expression of the Dynamics of the Observer in the System Coordinates

Chapter 8
Motivation and Problem Statement

Parts I–II have shown that it is possible, under certain conditions, to build an observer for a nonlinear system by transforming its dynamics into a favorable form for which a global observer is known. It follows that the dynamics of the system and of the observer are not expressed in the same coordinates and often evolve in spaces of different dimensions. It is therefore necessary to invert the injective transformation $T : \mathbb{R}^{d_x} \to \mathbb{R}^{d_\xi}$ (or more generally $T : \mathbb{R}^{d_x} \times \mathbb{R} \to \mathbb{R}^{d_\xi}$), not only to deduce $\hat{x}$ from $\hat{\xi}$, but also sometimes even to define the observer dynamics (for instance, in the high-gain framework, see below). But even if T is stationary, this inversion can be difficult in practice, mostly when an explicit expression for a global inverse is not available. Indeed, in this case, inversion usually relies on the resolution of a minimization problem of the type

$$\hat{x} = \min_{x \in \mathscr{X}} |T(x) - \hat{\xi}|$$

with a heavy computational cost.

In the case where T is a diffeomorphism on an open set $\mathscr{S}$ containing $\mathscr{X}$, one may hope to avoid this minimization by implementing the observer directly in the x-coordinates with

$$\dot{\hat{x}} = \left(\frac{dT}{dx}(\hat{x}) \right)^{-1} \mathscr{F}(T(\hat{x}), u, y) \tag{8.1}$$

This is done, for instance, in [3, 6, 11]. Note that even in this apparently simple case, this observer must be treated carefully, because although x stays in $\mathscr{S}$ where the Jacobian is invertible, there is no guarantee that $\hat{x}$ will, in particular during transients behaviors where peaking can occur. This problem will be treated later. For the moment, observe that this method does not extend easily to the more common case where $d_\xi > d_x$; i.e., the Jacobian is rectangular, and T is at best an injective immersion. This situation appears in a lot of applications, from walking robots [10], to aeronautics [8], or biochemistry [14], or micro-robotics [15, 16].

P. Bernard, *Observer Design for Nonlinear Systems*,
Lecture Notes in Control and Information Sciences 479,
https://doi.org/10.1007/978-3-030-11146-5_8

Some methods have recently been proposed to avoid solving the minimization problem:

- By using Newton-like or gradient-like algorithms. For instance, to the observer in the ξ coordinates
$$\dot{\hat{\xi}} = \mathscr{F}(\hat{\xi}, u, y),$$
we could add dynamics of the type
$$\dot{\hat{x}} = \mu K(\hat{x})(\hat{\xi} - T(\hat{x}))$$
where $K(x) = \frac{dT}{dx}(x)^\top$ or $K(x) = \left(\frac{dT}{dx}(x)^\top \frac{dT}{dx}(x)\right)^{-1} \frac{dT}{dx}(x)^\top$ as proposed in [12] for high-gain observers. But the convergence of the obtained observer is only local and practical.
- By continuation algorithms which "follow" the minimum of $x \mapsto |T(x) - \hat{\xi}|^2$, under a convexity assumption like in [9], but the convergence of the obtained observer is a priori only local and restricted to high-gain designs and triangular forms.
- When T is an injective immersion, by extending T into a diffeomorphism T_e and implementing the observer in the x-coordinates (8.1) using the extended transformation T_e instead of T, like in [2, 4, 5]. The advantage of this method is that it can provide global convergence and be applied to any observer, but the computation of the extension is not always straightforward.

Since the latter method applies to any type of observer, can be proved to provide global convergence, and tackles the important question of the choice of the observer coordinates, it belongs very well to the framework of this book. That is why we detail it in this part. Note that this method being somehow based on the "inversion" of the Jacobian, the injectivity of T no longer is sufficient, and we ask for an immersion, i.e., that the Jacobian be full-rank.

In this chapter, we first motivate and introduce this problem through examples and give a first sufficient condition to solve this problem in the case of a stationary transformation. The remaining Chaps. 9–11 will show how to satisfy this condition. Besides, the possible extension of those results to the case where the transformation is time-varying will be studied in Chap. 11 mainly through an example coming from an application.

8.1 Examples

8.1.1 Oscillator with Unknown Frequency

To motivate the problem we shall tackle in this part of the book, we consider a harmonic oscillator with unknown frequency with dynamics

$$\begin{cases} \dot{x}_1 = x_2 \\ \dot{x}_2 = -x_1 x_3 \\ \dot{x}_3 = 0 \end{cases} \quad , \quad y = x_1 \tag{8.2}$$

with state $x = (x_1, x_2, x_3)$ in $\mathbb{R}^2 \times \mathbb{R}_{>0}$ and measurement y. We are interested in estimating the state x of this system from the only knowledge of the function $t \mapsto y(t) = X_1(x, t)$.

For any solution with initial condition $x_1 = x_2 = 0$, y does not give any information on x_3. We thus restrict our attention to solutions evolving in

$$\mathscr{X} \subset \left\{ x \in \mathbb{R}^3 \; : \; x_1^2 + x_2^2 \in \left]\frac{1}{r}, r\right[\; , \; x_3 \in]0, r[\right\} , \tag{8.3}$$

where r is some arbitrary strictly positive real number. Note also that System (8.2) is strongly differentially observable of order 4 on

$$\mathscr{S} = \left(\mathbb{R}^2 \setminus \{(0,0)\}\right) \times \mathbb{R}_{>0}$$

containing $\mathscr{X}$, namely $\mathbf{H}_4$ defined by

$$\mathbf{H}_4(x) = \begin{pmatrix} h(x) \\ L_f h(x) \\ L_f^2 h(x) \\ L_f^3 h(x) \end{pmatrix} = \begin{pmatrix} x_1 \\ x_2 \\ -x_1 x_3 \\ -x_2 x_3 \end{pmatrix}$$

is an injective immersion on $\mathscr{S}$.

8.1.1.1 High-Gain Design

According to Theorem 7.1 and Remark 7.1, we know that T defined by

$$T(x) = \mathbf{H}_4(x) = (x_1, x_2, -x_1 x_3, -x_2 x_3) \tag{8.4}$$

transforms System (8.2) into a phase-variable form of dimension 4 for which a high-gain observer can be designed:

$$\dot{\hat{\xi}} = \mathscr{F}(\hat{\xi}, y) = \begin{pmatrix} 0 & 1 & 0 & 0 \\ 0 & 0 & 1 & 0 \\ 0 & 0 & 0 & 1 \\ 0 & 0 & 0 & 0 \end{pmatrix} \hat{\xi} + \begin{pmatrix} 0 \\ 0 \\ 0 \\ \Phi_4(\hat{\xi}) \end{pmatrix} + \begin{pmatrix} L k_1 \\ L^2 k_2 \\ L^3 k_3 \\ L^4 k_4 \end{pmatrix} [y - \hat{\xi}_1] , \tag{8.5}$$

where Φ_4 is defined by[1]

---

[1]The saturation function is defined by $\mathsf{sat}_M(s) = \min\{M, \max\{s, -M\}\}$.

$$\Phi_4(\xi) = \mathsf{sat}_{r^3}(L_f^4 h(\mathscr{T}(\xi)))$$

with $\mathscr{T}$ any locally Lipschitz function defined on $\mathbb{R}^4$ verifying

$$\mathscr{T}(\mathbf{H}_4(x)) = x \quad \forall x \in \mathscr{X} \ ,$$

r^3 may be replaced by any bound of $L_f^4 h$ on $\mathscr{X}$, and L is a sufficiently large strictly positive number depending on the Lipschitz constant of Φ_4, namely on the choice of $\mathscr{T}$ and r. Wanting to highlight the role of the computation of the left inverse $\mathscr{T}$, we get in fact a "raw" observer with dynamics

$$\dot{\hat{\xi}} \ = \ \varphi(\hat{\xi}, \hat{x}, y) \ = \ \begin{pmatrix} 0 & 1 & 0 & 0 \\ 0 & 0 & 1 & 0 \\ 0 & 0 & 0 & 1 \\ 0 & 0 & 0 & 0 \end{pmatrix} \hat{\xi} + \begin{pmatrix} 0 \\ 0 \\ 0 \\ \mathsf{sat}_{r^3}(\hat{x}_1 \hat{x}_3^2) \end{pmatrix} + \begin{pmatrix} \ell k_1 \\ \ell^2 k_2 \\ \ell^3 k_3 \\ \ell^4 k_4 \end{pmatrix} [y - \hat{\xi}_1] \quad , \quad \hat{x} = \mathscr{T}(\hat{\xi}) \ . \tag{8.6}$$

We deduce that the computation of the function $\mathscr{T}$ (whose existence is guaranteed by the theorem) is crucial in the implementation of this observer, of course to deduce $\hat{x}$ from $\hat{\xi}$ but also to define the dynamics of the observer itself.

Although in this example an explicit and global expression[2] for $\mathscr{T}$ can easily be found due to the simplicity of the transformation $T = \mathbf{H}_4$, it is not always the case in high-gain designs for more complex applications [10, 14]. An example will be given below in Sect. 8.1.2.

8.1.1.2 Luenberger Design

Instead of a high-gain observer design as above, we may use a nonlinear Luenberger design. As explained in Sect. 6.2, the idea is to find a transformation into a Hurwitz form of the type

$$\dot{\xi} \ = \ A\,\xi \ + \ B\,y \ ,$$

with ξ in $\mathbb{R}^{d_\xi}$, A a Hurwitz matrix, and (A, B) a controllable pair. Indeed, this system admits as global observer

$$\dot{\hat{\xi}} \ = \ \varphi(\hat{\xi}, y) \ = \ A\,\hat{\xi} \ + \ B\,y \ . \tag{8.7}$$

Since the dynamics (8.2) are linear in (x_1, x_2), we can look for a transformation depending linearly in (x_1, x_2). Straightforward computations give:

$$T(x) \ = \ -(A^2 + x_3 I)^{-1}[ABx_1 + Bx_2] \ . \tag{8.8}$$

[2]For instance, we can take $\mathscr{T}(\xi) = \left(\xi_1\ ,\ \xi_2\ ,\ -\frac{\xi_1\xi_3+\xi_4\xi_2}{\max\{\xi_1^2+\xi_2^2,\frac{1}{r^2}\}}\right)$.

In particular, for a diagonal matrix $A = \mathsf{diag}(-\lambda_1, \dots, -\lambda_{d_\xi})$ with $\lambda_i > 0$, and $B = (1, \dots, 1)^\top$, this gives

$$T_i(x) = \frac{\lambda_i x_1 - x_2}{\lambda_i^2 + x_3} \tag{8.9}$$

for i in $\{1, \dots, d_\xi\}$. It is shown in [13] that T is injective on $\mathscr{S}$ if $d_\xi \geq 4$ for any distinct λ_i's in $(0, +\infty)$. More precisely, it is Lipschitz injective on any compact subset $\mathscr{X}$ of $\mathscr{S}$; i.e., there exists $\mathfrak{a}$ such that

$$|x_a - x_b| \leq \mathfrak{a}|T(x_a) - T(x_b)| \qquad \forall (x_a, x_b) \in \mathscr{X}^2 .$$

From this,[3] we get that T is actually an injective immersion on $\mathscr{S}$. This is consistent with [1, Theorem 4] and the fact that the order of strong differentiability of this system is 4.

Thus, since the trajectories of the system remain bounded, applying Corollary 1.1, there exists an observer for System (8.2) which is given by (8.7) and any continuous function $\mathscr{T}$ satisfying

$$\mathscr{T}(T(x)) = x \quad \forall x \in \mathscr{X} .$$

However, it is difficult to find an explicit expression of $\mathscr{T}$, so that for this design, we would have to solve online

$$\hat{x} = \mathscr{T}(\hat{\xi}) = \text{Argmin}_{\hat{x}} \left|\hat{\xi} - T(\hat{x})\right|^2 .$$

or use more local techniques as mentioned in the introduction. Note that a difference with the high-gain observer above is that $\hat{x}$ is not involved in (8.7); i.e., the observer dynamics do not depend on $\mathscr{T}$.

8.1.2 Bioreactor

Consider a bioreactor model as in [14]

$$\begin{cases} \dot{x}_1 = \mu(x_2)x_1 \\ \dot{x}_2 = -k\mu(x_2)x_1 \end{cases} , \quad y = x_1 \tag{8.10}$$

on $\Omega = \mathbb{R}_{>0} \times \mathbb{R}_{>0}$, where x_1 (resp. x_2) is the biomass (resp. substrate) concentration, k is a positive constant, and μ is a nonnegative smooth function such that

[3] Indeed, consider any x in $\mathscr{S}$ and $\mathscr{V}$ an open neighborhood of x such that $\mathtt{cl}(\mathscr{V})$ is contained in $\mathscr{S}$. According to the Lipschitz injectivity of T on $\mathtt{cl}(\mathscr{V})$, there exists $\mathfrak{a}$ such that for all v in $\mathbb{R}^3$ and for all h in $\mathbb{R}$ such that $x + hv$ is in $\mathscr{V}$, $|v| \leq \mathfrak{a}\frac{|T(x+hv)-T(x)|}{|h|}$ and thus by taking h to zero, $|v| \leq \mathfrak{a}\left|\frac{\partial T}{\partial x}(x)v\right|$ which means that $\dfrac{\partial T}{\partial x}(x)$ is full-rank.

$\mu(0) = 0$ and μ is non-monotonic. Examples of such a map μ are of the form

$$\mu(s) = k_1 s \left(1 - \frac{s}{k_2}\right) , \tag{8.11}$$

or

$$\mu(s) = \frac{s}{k_1 + s + k_2 s^2} , \tag{8.12}$$

where the decreasing phase rendering μ non-monotonic models an inhibiting effect of the reaction. For a bounded subset of initial conditions $\mathcal{X}_0 \subset \Omega$ of interest, there exists a compact set $\mathcal{X} \subset \Omega$ such that any trajectory initialized in $\mathcal{X}_0$ stays in $\mathcal{X}$.

Assume we want to do a high-gain design as in [14]. Since $x \mapsto (h(x), L_f h(x))$ is not injective, we take

$$\begin{aligned} T(x) &= (h(x), L_f h(x), L_f h^2(x)) \\ &= (x_1, \mu(x_2)x_1, \mu(x_2)x_1(-k\mu'(x_2)x_1 + \mu(x_2))) \end{aligned} \tag{8.13}$$

which is typically an injective immersion on an open subset $\mathcal{S}$ of Ω containing the set $\mathcal{X}$ where the solutions evolve. In other words, the system is strongly differentially observable of order 3 on the domain of interest. Similar to the oscillator with unknown frequency from Sect. 8.1.1.1, a high-gain design leads us to the following raw observer

$$\dot{\hat{\xi}} = \varphi(\hat{\xi}, \hat{x}, y) = \begin{pmatrix} 0 & 1 & 0 \\ 0 & 0 & 1 \\ 0 & 0 & 0 \end{pmatrix} \hat{\xi} + \begin{pmatrix} 0 \\ 0 \\ \mathsf{sat}_R(L_f^3 h(\hat{x})) \end{pmatrix} + \begin{pmatrix} \ell k_1 \\ \ell^2 k_2 \\ \ell^3 k_3 \end{pmatrix} [y - \hat{\xi}_1] \quad , \quad \hat{x} = \mathcal{T}(\hat{\xi}) , \tag{8.14}$$

where R is a bound of $L_f^3 h$ on $\mathcal{X}$ and $\mathcal{T}$ verifies

$$\mathcal{T}(T(x)) = x \quad \forall x \in \mathcal{X} .$$

Again, inverting T, namely computing $\mathcal{T}$, is challenging and no explicit expression is available. In [14], this is done by building a globally Lipschitz extension of the left inverse of T. Here, we wonder how those difficult steps can be avoided.

8.1.3 General Idea

In the following, we detail the methodology proposed in [2, 5] to write the dynamics of the given observers (8.6), (8.7), or (8.14) directly in the x-coordinates.[4]

In the example above, pulling the observer dynamics from the ξ-coordinates back to the x-coordinates appears impossible since x has dimension 3, whereas ξ has

[4]We will also refer to the x-coordinates as the "given coordinates" because they are chosen by the user to describe the model dynamics.

dimension 4. This difficulty is overcome by adding one component, say w, to x. Then, the dynamics of $(\hat{x}, \hat{w})$ can be obtained from those of $\hat{\xi}$ if we have a diffeomorphism $(x, w) \mapsto \xi = T_e(x, w)$ "augmenting" the function $x \mapsto T(x)$ given in (8.4), (8.8), or (8.13). We will see in Chap. 9 how this can be done by complementing the full-rank rectangular Jacobian T into an invertible matrix. Unfortunately, in doing so (and as mentioned in the introduction), the obtained diffeomorphism is rarely defined everywhere and one has no guarantee that the trajectory of $(\hat{x}, \hat{w})$ remains in the domain of definition of the diffeomorphism. We will see in Chap. 10 how this new problem can be overcome via extending the image of the diffeomorphism. The key point here is that the given observer dynamics (8.6) or (8.7) remain unchanged in the ξ-coordinates. This differs from other techniques as proposed in [3, 11], which require extra assumptions such as convexity to preserve the convergence property.

8.2 Problem Statement

8.2.1 *Starting Point*

We consider[5] a given system with dynamics

$$\dot{x} = f(x, u) \quad , \qquad y = h(x, u) \,, \tag{8.15}$$

with x in $\mathbb{R}^{d_x}$, u a function in $\mathscr{U}$ with values in $U \subset \mathbb{R}^{d_u}$, and y in $\mathbb{R}^{d_y}$. The observation problem is to construct a dynamical system with input y and output $\hat{x}$, supposed to be an estimate of the system state x as long as the latter is in a specific set of interest denoted $\mathscr{X} \subseteq \mathbb{R}^{d_x}$. As starting point here, we assume this problem is (formally) already solved but with maybe some implementation issues such as finding an expression of $\mathscr{T}$. More precisely,

Assumption 8.1 (*Converging observer in the ξ-coordinates*) There exist an open subset $\mathscr{S}$ of $\mathbb{R}^{d_x}$, a subset $\mathscr{X}$ of $\mathscr{S}$, a C^1 injective immersion $T : \mathscr{S} \to \mathbb{R}^{d_\xi}$, and a set $\varphi\mathscr{T}$ of pairs $(\varphi, \mathscr{T})$ of functions such that:

- $\mathscr{T} : \mathbb{R}^{d_\xi} \to \mathbb{R}^{d_x}$ is a left inverse of T on $T(\mathscr{X})$, i.e.,

$$\mathscr{T}(T(x)) = x \qquad \forall x \in \mathscr{X} \tag{8.16}$$

- For any u in $\mathscr{U}$ and any x_0 in $\mathscr{X}_0$ such that $\sigma^+(x_0, u) = +\infty$, the solution $X(x_0; t; u)$ of (8.15) remains in $\mathscr{X}$ for t in $[0, +\infty)$.
- For any u in $\mathscr{U}$, any x_0 in $\mathscr{X}_0$ such that $\sigma^+(x_0, u) = +\infty$, and any $\hat{\xi}_0$ in $\mathbb{R}^{d_\xi}$, any solution $(X(x_0; t; u), \hat{\Xi}((x_0, \hat{\xi}_0); t; u))$ of the cascade system:

[5]Texts of Sect. 8.2 are reproduced from [5] with permission from SIAM.

$$\dot{x} = f(x,u) \quad , \quad y = h(x,u) \quad , \quad \dot{\hat{\xi}} = \varphi(\hat{\xi},\hat{x},u,y) \quad , \quad \hat{x} = \mathscr{T}(\hat{\xi}) \, , \tag{8.17}$$

initialized at $(x_0, \hat{\xi}_0)$ and under the input u, is also defined on $[0, +\infty)$ and satisfies:

$$\lim_{t\to+\infty} \left| \Xi((x_0,\hat{\xi}_0);t;u) - T(X(x_0;t;u)) \right| = 0 \, . \tag{8.18}$$

Remark 8.1

1. The convergence property given by (8.18) is in the observer state space only. Property (8.16) is a necessary condition for this convergence to be transferred from the observer state space to the system state space. But as explained in Chap. 1, we may need the injectivity of T to be uniform in space, or equivalently $\mathscr{T}$ to be uniformly continuous on $\mathbb{R}^{d_\xi}$, in order to conclude about a possible convergence in the x-coordinates. In that case, the couple $(\mathscr{F}, \mathscr{T})$ with

   $$\mathscr{F}(\xi,u,y) = \varphi(\xi, \mathscr{T}(\xi), u, y)$$

   is an observer for System (8.15) initialized in $\mathscr{X}_0$. Note that as in Corollary 1.1, this is achieved without further assumption in the case where $\mathscr{X}$ is bounded.
2. The reason why we make φ depend on $\hat{x}$, instead of simply taking $\mathscr{F}(\xi,u,y)$ as before, is that most of the time, and especially in a high-gain design (see (8.6) or (8.14)), when expressing $\mathscr{F}$ as a function of ξ, we replace x by $\mathscr{T}(\xi)$. Since we want here to avoid the computation of $\mathscr{T}$, we make this dependence explicit in φ.
3. The need for pairing φ and $\mathscr{T}$ comes from this dependence because it may imply to change φ whenever we change $\mathscr{T}$. In the high-gain approach, for instance, the high gain L must be adapted to the Lipschitz constant of the nonlinearity entering on the last line, and therefore to the Lipschitz constant of $\mathscr{T}$, since this nonlinearity is a saturated version of $L_f^m h \circ \mathscr{T}$. Hence, when $\mathscr{X}$ is bounded, φ with a sufficiently large gain can be paired with any locally Lipschitz function $\mathscr{T}$. On another hand, if, as in (8.7), φ does not depend on $\hat{x}$, then it can be paired with any $\mathscr{T}$.

Example 8.1 For System (8.2), $\mathscr{X}$ given in (8.3) being bounded, a set $\varphi\mathscr{T}$ satisfying Assumption 8.1 is made of pairs of

- A locally Lipschitz function $\mathscr{T}$ satisfying

  $$x = \mathscr{T}(x_1, x_2, -x_1x_3, -x_2x_3) \qquad \forall x \in \mathscr{X} \tag{8.19}$$

  and the function φ defined in (8.6), with L adapted to the Lipschitz constant of $\mathscr{T}$ around $T(\mathscr{X})$, if T is defined by (8.4);
- Or a continuous function $\mathscr{T}$ satisfying

$$x = \mathscr{T}\left(\frac{\lambda_1 x_1 - x_2}{\lambda_1^2 + x_3}, \frac{\lambda_2 x_1 - x_2}{\lambda_2^2 + x_3}, \frac{\lambda_3 x_1 - x_2}{\lambda_3^2 + x_3}, \frac{\lambda_4 x_1 - x_2}{\lambda_4^2 + x_3}\right) \qquad \forall x \in \mathscr{X} \tag{8.20}$$

and the function φ defined in (8.7) if T is defined by (8.9). ▲

Similarly, for System (8.10), a set $\varphi\mathscr{T}$ satisfying Assumption 8.1 is made of pairs of a locally Lipschitz function $\mathscr{T}$ satisfying

$$x = \mathscr{T}\Big(x_1\,,\ \mu(x_2)x_1\,,\ \mu(x_2)x_1(-k\mu'(x_2)x_1 + \mu(x_2)\Big) \qquad \forall x \in \mathscr{X} \tag{8.21}$$

and the function φ defined in (8.14), with L adapted to the Lipschitz constant of $\mathscr{T}$ around $T(\mathscr{X})$.

Although the problem of observer design seems already solved under Assumption 8.1, we have seen that it can be challenging to find a left inverse $\mathscr{T}$ of T. In the following, we consider that the function T and the set $\varphi\mathscr{T}$ are given and we aim at avoiding the left inversion of T by expressing the observer for x in the, maybe augmented, x-coordinates.

8.2.2 A Sufficient Condition Allowing the Expression of the Observer in the Given x-Coordinates

For the simpler case where the raw observer state $\hat{\xi}$ has the same dimension as the system state x, i.e., $d_x = d_\xi$, T in Assumption 8.1 is a diffeomorphism on $\mathscr{S}$ and we can express the observer in the given x-coordinates as:

$$\dot{\hat{x}} = \left(\frac{\partial T}{\partial x}(\hat{x})\right)^{-1} \varphi(T(\hat{x}), \hat{x}, u, y) \tag{8.22}$$

which requires a Jacobian inversion only. However, although, by assumption, the system trajectories remain in $\mathscr{S}$ where the Jacobian is invertible, we have no guarantee the ones of the observer do. Therefore, to obtain convergence and completeness of solutions, we must find means to ensure the estimate $\hat{x}$ does not leave the set $\mathscr{S}$ or equivalently that $T(\hat{x})$ remains in the image set $T(\mathscr{S})$. Some authors have proposed to use saturations/projections [11] or to modify the dynamics of the observer [3], namely φ, in order to force the state to remain in $T(\mathscr{S})$. However, this cannot be done lightly because it could destroy the observer convergence, and in both [3, 11], some convexity assumptions on the set $T(\mathscr{S})$ are made, which unfortunately are rarely satisfied. The approach proposed in [4] is rather to keep the observer dynamics φ intact, but change the map T instead. Indeed, observe that this problem disappears if $T(\mathscr{S})$ is the whole space $\mathbb{R}^{d_\xi}$. The idea is therefore to modify T outside of $\mathscr{X}$ (where the solutions evolve) in order to get $T(\mathscr{S}) = \mathbb{R}^{d_\xi}$, namely extend the image of the diffeomorphism T.

In the more complex situation where $d_\xi > d_x$, T is only an injective immersion, and $T(\mathscr{S})$ is a manifold of dimension d_x in a space of dimension d_ξ. In order to use the same ideas as described in the previous paragraph, it was first proposed in [2] to "transform" the map T into a diffeomorphism. This is done by augmenting the given x-coordinates in $\mathbb{R}^{d_x}$ with extra ones, say w, in $\mathbb{R}^{d_\xi - d_x}$ and correspondingly augmenting the given injective immersion T into a diffeomorphism $T_e : \mathcal{S}_a \to \mathbb{R}^{d_\xi}$, where $\mathcal{S}_a$ is an open subset of $\mathbb{R}^{d_\xi}$, which "augments" $\mathcal{S}$; i.e., its Cartesian projection on $\mathbb{R}^{d_x}$ is contained in $\mathcal{S}$ and contains $\mathtt{cl}(\mathcal{X})$. The idea is again to keep the converging observer dynamics, but change the way the estimate $\hat{\xi}$ is associated to $\hat{x}$.

To help us find such an appropriate augmentation, the following sufficient condition is given in [5].

Theorem 8.1 ([5]) *Assume Assumption 8.1 holds and $\mathcal{X}$ is bounded. Assume also the existence of an open subset $\mathcal{S}_a$ of $\mathbb{R}^{d_\xi}$ containing $\mathtt{cl}(\mathcal{X} \times \{0\})$ and of a diffeomorphism $T_e : \mathcal{S}_a \to \mathbb{R}^{d_\xi}$ satisfying*

$$T_e(x,0) \;=\; T(x) \qquad \forall x \in \mathcal{X} \tag{8.23}$$

and

$$T_e(\mathcal{S}_a) = \mathbb{R}^{d_\xi} \ , \tag{8.24}$$

and such that, with $\mathcal{T}_{ex}$ denoting the x-component of T_e^{-1}, there exists a function φ such that the pair $(\varphi, \mathcal{T}_{ex})$ is in the set $\varphi\mathcal{T}$ given by Assumption 8.1.

Under these conditions, for any u in $\mathcal{U}$ and any x_0 in $\mathcal{X}_0$ such that $\sigma^+(x_0,u) = +\infty$, any solution $(X(x_0;t;u), \hat{X}(x_0,\hat{x}_0,\hat{w}_0;t;u), \hat{W}(x_0,\hat{x}_0,\hat{w}_0;t,u))$, with initial condition $(\hat{x}_0,\hat{w}_0)$ in $\mathcal{S}_a$, of the cascade of System (8.15) with the observer

$$\dot{\overbrace{\begin{bmatrix}\hat{x}\\ \hat{w}\end{bmatrix}}} = \left(\frac{\partial T_e}{\partial(\hat{x},\hat{w})}(\hat{x},\hat{w})\right)^{-1} \varphi(T_e(\hat{x},\hat{w}),\hat{x},u,y) \tag{8.25}$$

is also defined on $[0,+\infty)$ and satisfies:

$$\lim_{t\to+\infty} \left|\hat{W}(x_0,\hat{x}_0,\hat{w}_0;t;u)\right| + \left|X(x_0;t;u) - \hat{X}(x_0,\hat{x}_0,\hat{w}_0;t;u)\right| = 0 \,. \tag{8.26}$$

In other words, System (8.25) is an observer in the given coordinates[6] *for System (8.15) initialized in $\mathcal{X}_0$.*

The key point in the observer (8.25) is that, instead of left inverting the function T via $\mathcal{T}$ as in (8.16), we invert only a matrix, exactly as in (8.22). The proof is instructive to see where each assumption plays a role.

Proof Take u in $\mathcal{U}$ and $(x_0,(\hat{x}_0,\hat{w}_0))$ in $\mathcal{X}_0 \times \mathcal{S}_a$ such that $\sigma^+(x_0,u) = +\infty$. $X(x_0;t;u)$ remains in $\mathcal{X}$ for t in $[0,+\infty)$ by assumption. Let $[0,\bar{t}[$ be the right maximal interval of definition of the solution $(X(x_0,t), \hat{X}(x_0,\hat{x}_0,\hat{w}_0;t;u), \hat{W}(x_0,\hat{x}_0,\hat{w}_0;t;u))$ when considered with values in $\mathcal{X} \times \mathcal{S}_a$. Assume for the time being $\bar{t}$ is finite. Then, when t goes to $\bar{t}$, either $(\hat{X}(x_0,\hat{x}_0,\hat{w}_0;t;u), \hat{W}(x_0,\hat{x}_0,\hat{w}_0;t;u))$ goes to infinity or to the boundary of $\mathcal{S}_a$. By construction $t \mapsto \Xi(t) := T_e\left(\hat{X}(\hat{x}_0,\hat{w}_0;t;u), \hat{W}(\hat{x}_0,\hat{w}_0;t;u)\right)$ is a solution of (8.17) on $[0,\bar{t}[$ with $\mathcal{T} = \mathcal{T}_{ex}$.

[6] See Definition 1.1.

From Assumption 8.1 and since $(\varphi, \mathscr{T}_{ex})$ is in $\wp\mathscr{T}$, it can be extended as a solution defined on $[0, +\infty[$ when considered with values in $\mathbb{R}^{d_\xi} = T_e(\mathscr{S}_a)$. This implies that $\Xi(\bar{t})$ is well defined in $\mathbb{R}^{d_\xi}$. Since, with (8.24), the inverse T_e^{-1} of T_e is a diffeomorphism defined on $\mathbb{R}^{d_\xi}$, we obtain $\lim_{t \to \bar{t}} \left(\hat{X}(\hat{x}_0, \hat{w}_0; t; u), \hat{W}(\hat{x}_0, \hat{w}_0; t; u)\right) = T_e^{-1}(\Xi(\bar{t}))$, which is an interior point of $T_e^{-1}(\mathbb{R}^{d_\xi}) = \mathscr{S}_a$. This point being neither a boundary point nor at infinity, we have a contradiction. It follows that $\bar{t}$ is infinite.

Finally, with assumption 8.1, we have

$$\lim_{t \to +\infty} \left| T_e\left(\hat{X}(\hat{x}_0, \hat{w}_0; t; u), \hat{W}(\hat{x}_0, \hat{w}_0; t; u)\right) - T(X(x_0; t; u)) \right| = 0 .$$

Since $X(x_0; t; u)$ remains in $\mathscr{X}$, $T(X(x_0; t; u))$ equals $T_e(X(x_0; t; u), 0)$ from (8.23) and remains in the compact set $T(\mathtt{cl}(\mathscr{X}))$. So there exists a compact subset $\mathscr{C}$ of $\mathbb{R}^{d_\xi}$ and a time $t_{\mathscr{C}}$ such that $T_e\left(\hat{X}(\hat{x}_0, \hat{w}_0; t; u), \hat{W}(\hat{x}_0, \hat{w}_0; t; u)\right)$ is in $\mathscr{C}$ for all $t > t_{\mathscr{C}}$. Since T_e is a diffeomorphism, its inverse T_e^{-1} is Lipschitz on the compact set $\mathscr{C}$. This implies (8.26). □

With Theorem 8.1, we are left with finding a diffeomorphism T_e satisfying the conditions listed in the statement:

- Equation (8.23) is about the fact that T_e is an augmentation, with adding coordinates, of the given injective immersion T. It motivates the following problem.

Problem 8.1 (*Immersion augmentation into a diffeomorphism*) Given a set $\mathscr{X}$, an open subset $\mathscr{S}$ of $\mathbb{R}^{d_x}$ containing $\mathtt{cl}(\mathscr{X})$, and an injective immersion $T : \mathscr{S} \to T(\mathscr{S}) \subset \mathbb{R}^{d_\xi}$, the pair $(T_a, \mathscr{S}_a)$ is said to solve the problem of *immersion augmentation into a diffeomorphism* if $\mathscr{S}_a$ is an open subset of $\mathbb{R}^{d_\xi}$ containing $\mathtt{cl}(\mathscr{X} \times \{0\})$ and $T_a : \mathscr{S}_a \to T_a(\mathscr{S}_a) \subset \mathbb{R}^{d_\xi}$ is a diffeomorphism satisfying

$$T_a(x, 0) = T(x) \qquad \forall x \in \mathscr{X} .$$

We will see in Chap. 9 conditions under which Problem 8.1 can be solved via complementing a full-column rank Jacobian of T into an invertible matrix, i.e., via what is called in [4] Jacobian complementation.

- The condition expressed in (8.24) is about the fact that T_e is surjective onto $\mathbb{R}^{d_\xi}$. This motivates us to introduce the following problem

Problem 8.2 (*Surjective diffeomorphism extension*) Given an open subset $\mathscr{S}_a$ of $\mathbb{R}^{d_\xi}$, a compact subset K of $\mathscr{S}_a$, and a diffeomorphism T_a: $\mathscr{S}_a \to \mathbb{R}^{d_\xi}$, the diffeomorphism $T_e : \mathscr{S}_a \to \mathbb{R}^{d_\xi}$ is said to solve the *surjective diffeomorphism extension problem* if it satisfies

$$T_e(\mathscr{S}_a) = \mathbb{R}^{d_\xi} \quad , \qquad T_e(x, w) = T_a(x, w) \quad \forall (x, w) \in K.$$

This Problem 8.2 will be addressed in Chap. 10.

When Assumption 8.1 holds and $\mathcal{X}$ is bounded, by successively solving Problem 8.1 and Problem 8.2 with $\mathtt{cl}(\mathcal{X} \times \{0\}) \subset K \subset \mathcal{S}_a$, we get a diffeomorphism T_e guaranteed to satisfy all the conditions of Theorem 8.1 except maybe the fact that the pair $(\varphi, \mathcal{T}_{ex})$ is in $\varphi\mathcal{T}$. Fortunately, pairing a function φ with a function $\mathcal{T}_{ex}$ obtained from a left inverse of T_e is not as difficult as it seems, at least for general purpose observer designs such as high-gain observers or nonlinear Luenberger observers. Indeed, we have already observed in item 3 of Remark 8.1 that if, as for nonlinear Luenberger observers, or linearization by output feedback, there is a pair $(\varphi, \mathcal{T})$ in the set $\varphi\mathcal{T}$ such that φ does not depend on $\mathcal{T}$, then we can associate this φ to any $\mathcal{T}_{ex}$. Also, for high-gain observers, we need only that $\mathcal{T}_{ex}$ be locally Lipschitz, which is for free because it is continuously differentiable by definition of a diffeomorphism. We conclude from all this that the problem reduces to solving Problems 8.1 and 8.2.

Actually, although Theorem 8.1 suggests to solve sequentially Problem 8.1 and then Problem 8.2, one may also wonder if it is possible to build directly T_e from T. This is done in the following section on an example thanks to a graphical method.

8.3 Direct Construction of the Extended Diffeomorphism T_e?

Let us come back to the bioreactor model (8.10). Observe that its dynamics can be simplified by performing a change of time given by

$$\tau(t) = \int_0^t x_1(s)ds$$

which is strictly increasing and such that $\tau(0) = 0$ and $\lim_{t\to+\infty} \tau(t) = +\infty$ (since x_1 increases and is lower bounded by its positive initial condition). The trajectories $s \mapsto x_1(\tau^{-1}(s)), s \mapsto x_2(\tau^{-1}(s)), s \mapsto y(\tau^{-1}(s))$ in the new time frame follow[7] the dynamics

$$\begin{cases} \dot{x}_1 = \mu(x_2) \\ \dot{x}_2 = -k\mu(x_2) \end{cases}, \quad y = x_1 . \tag{8.27}$$

It is easy to check that finding an observer for system (8.27) in time s leads to an observer for (8.10) in time t by multiplying the observer dynamics by y.

System (8.27) is strongly differentially observable of order 3; namely, the map

$$T(x) = (x_1\,,\ \mu(x_2)\,,\ -k\mu(x_2)\mu'(x_2)) \tag{8.28}$$

is an injective immersion on Ω, whether μ be defined as (8.11) or (8.12). We can therefore design a high-gain observer of dimension 3 in the same way we did for

[7] We abusively denote in same way $t \mapsto x(t)$ and $s \mapsto x(\tau^{-1}(s))$, $t \mapsto y(t)$ and $s \mapsto y(\tau^{-1}(s))$, $\dot{x}$ and $\frac{dx}{ds}$ although those are in different time frames.

(8.10). Then, according to Theorem 8.1, those observer dynamics can be written in the x-coordinates if the map T can be extended into a surjective diffeomorphism.

The advantage of this map T compared to (8.13) is that x_1 and x_2 are totally decoupled and the part concerning x_1 is just the identity map. Therefore, it is enough to extend only the part in x_2; i.e., look for T_e of the form

$$T_e(x, w) = (x_1, T_{e,23}(x_2, w)) \tag{8.29}$$

where $T_{e,23} : \mathbb{R}^2 \to \mathbb{R}^2$ is a surjective diffeomorphism extension of the injective immersion $T_{23} : \mathbb{R}_{>0} \to \mathbb{R}^2$ defined by

$$T_{23}(x_2) = (\mu(x_2)\,,\ -k\mu(x_2)\mu'(x_2))\ .$$

We are going to build this map explicitly, thanks to graphical considerations in $\mathbb{R}^2$.

Consider $c > 0$ such that the component x_2 of the solutions is known to stay in $[0, c]$. We want to build a diffeomorphism $T_{e,23} : \mathbb{R}^2 \to \mathbb{R}^2$ such that

$$T_{e,23}(x_2, 0) = T_{23}(x_2) \quad \forall x_2 \in [0, c]\ ,$$

and such that its image covers the whole $\mathbb{R}^2$. Our starting point is thus $T_{23}([0, c])$ which is plotted on Fig. 8.1a. Notice that this is a manifold of dimension 1 in $\mathbb{R}^2$ which will be the level set of $T_{e,23}$ at $w = 0$. We now need to build the level sets $T_{e,23}([0, c], w)$ for $w \neq 0$ such that the whole space $\mathbb{R}^2$ is covered. Before that, it is important to note that the level sets must never cross each other for $T_{e,23}$ to be injective.

The first idea is to continuously transform $T_{23}([0, c])$ into a simpler "curve"; namely, use an homotopy with w as parameter. If the chosen target function for $w = 1$ is $x_2 \mapsto \overline{T}_{23}(x_2)$, one can take the convex combination

$$T_{e,23}(x_2, w) = (1 - w)T_{23}(x_2) + w\,\overline{T}_{23}(x_2) \quad , \quad (x_2, w) \in [0, c] \times [0, 1]\ . \tag{8.30}$$

For instance, let us say we want the level set at $w = 1$ to look like the parabola $\mathscr{P}$ plotted in Fig. 8.1b: In other words, we want to choose $\overline{T}_{23}(x_2)$ such that $\overline{T}_{23}([0, c]) = \mathscr{P}$. A possibility is to define $\overline{T}_{23}(x_2)$ as a projection[8] of $T_{23}(x_2)$ on $\mathscr{P}$. Using (8.30), this gives the level sets plotted on Fig. 8.1c.

Then, it turns out that due to the shape of the curve, the same formula (8.30) also works for $w \leq 0$; namely, it produces level sets that do not intersect, as shown on Fig. 8.1d. Therefore, we take

$$T_{e,23}(x_2, w) = (1 - w)T_{23}(x_2) + w\,\overline{T}_{23}(x_2) \quad , \quad (x_2, w) \in [0, c] \times (-\infty, 1]\ . \tag{8.31}$$

[8]For instance, choosing a center C inside the parabola, define the function α such that for any $\xi \in \mathbb{R}^2$ outside $\mathscr{P}$, $\alpha(\xi)$ is the unique positive number such that $C + \alpha(\xi)(\xi - C) \in \mathscr{P}$. Then, take $\overline{T}_{23}(x_2) = C + \alpha(T_{23}(x_2))(T_{23}(x_2) - C)$.

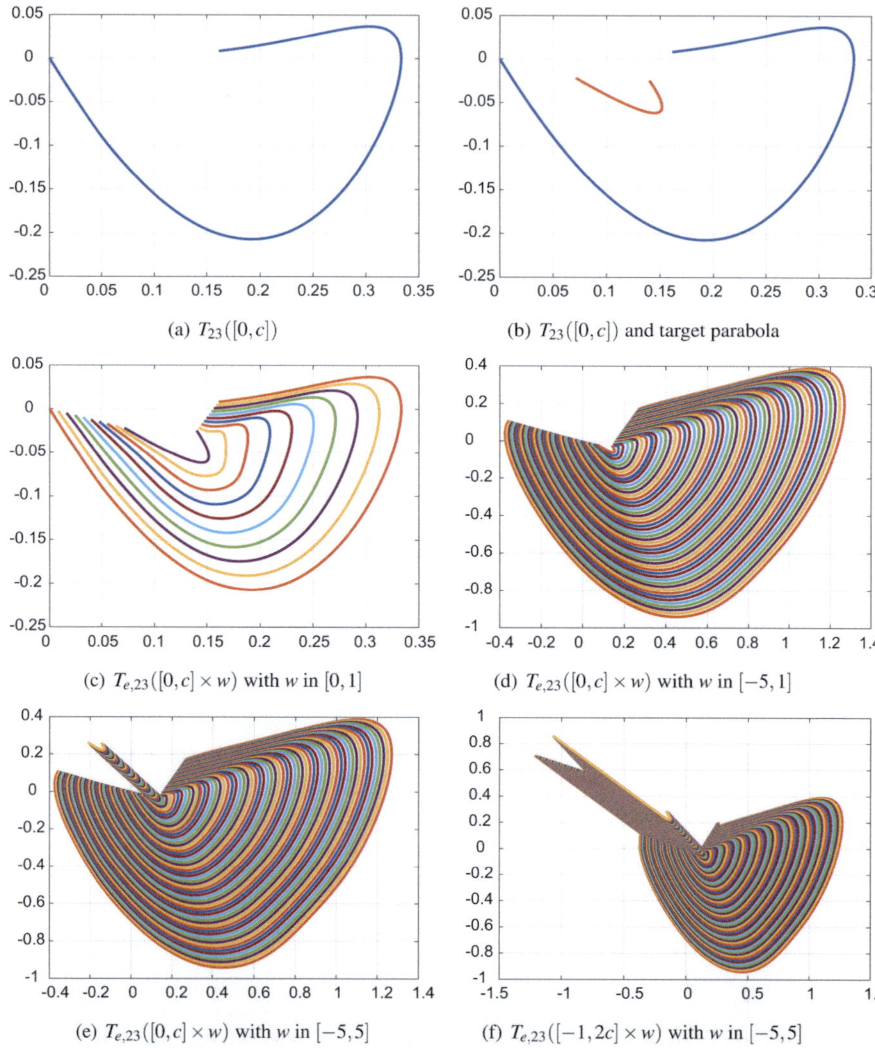

(a) $T_{23}([0,c])$

(b) $T_{23}([0,c])$ and target parabola

(c) $T_{e,23}([0,c]\times w)$ with w in $[0,1]$

(d) $T_{e,23}([0,c]\times w)$ with w in $[-5,1]$

(e) $T_{e,23}([0,c]\times w)$ with w in $[-5,5]$

(f) $T_{e,23}([-1,2c]\times w)$ with w in $[-5,5]$

Fig. 8.1 Construction of the extension $T_{e,23}$ of T_{23} with μ defined in (8.12), $k_1 = k_2 = 1$, and $c = 5$

Unfortunately, this is no longer possible for $w \geq 1$.

Actually, for $w \geq 1$, one can simply translate the parabola along its axis; namely, take

$$T_{23,e}(x_2, w) = \overline{T}_{23}(x_2) + (w-1)v \quad , \quad (x_2, w) \in [0, c] \times [1, +\infty) \, , \qquad (8.32)$$

with $v \in \mathbb{R}^2$ the direction of the translation. The result is shown on Fig. 8.1e.

At this point, we have an injective function defined on $[0, c] \times \mathbb{R}$. Outside of $[0, c]$, the expression of T_{23} cannot be kept because T_{23} tends to 0 when x_2 goes to infinity, thus making injectivity impossible. A simple solution consists in extending the parabola and T_{23} with lines of same slope and keeping the same expressions (8.31)–(8.32) outside of $[0, c]$ with T_{23} and $\overline{T}_{23}$ modified accordingly. This is illustrated on Fig. 8.1f: The level sets never intersect, and we obtain an injective and surjective map $T_{e,23} : \mathbb{R}^2 \to \mathbb{R}^2$.

Of course, the last step is to ensure the regularity of $T_{e,23}$, and some regularizations should be carried out around $x_2 = 0$, $x_2 = c$, and $w = 1$. In that way, we obtain a globally defined and surjective diffeomorphism which coincides with T_{23} on $[0, c] \times \{0\}$. Those properties are directly extended to T_e defined in (8.29). Therefore, we can write the high-gain observer in the x-coordinates using (8.25) and then come back to the initial time frame by multiplying them by y.

Unfortunately, the explicit global extension of T into a diffeomorphism that we just managed in this particular example cannot easily be generalized because it relies on graphical considerations in dimension 2 that would become challenging in dimension 3 and impossible in higher dimensions.

That is why, throughout Chaps. 9 and 10, we intend to solve Problems 8.1 and 8.2 generally. We will see how, step by step, we can express in the x-coordinates the high-gain observer for the harmonic oscillator with unknown frequency introduced in Sect. 8.1.1.1. We will also show that this approach enables to ensure completeness of solutions of the observer presented in [7] for a bioreactor. The various difficulties we shall encounter on this road will be discussed in Chap. 11. In particular, we shall see how they can be overcome thanks to a better choice of T and of the pair (φ, τ) given by Assumption 8.1. We will also see that the same tools apply to the Luenberger observer presented in Sect. 8.1.1.2 for the oscillator. Finally, in Chap. 11, we will try to extend this methodology to the case where the transformation is time-varying through a very practical application related to aircraft landing.

References

1. Andrieu, V., Praly, L.: On the existence of a Kazantzis-Kravaris/Luenberger observer. SIAM J. Control Optim. **45**(2), 432–456 (2006)
2. Andrieu, V., Eytard, J.B., Praly, L.: Dynamic extension without inversion for observers. In: IEEE Conference on Decision and Control, pp. 878–883 (2014)
3. Astolfi, D., Praly, L.: Output feedback stabilization for SISO nonlinear systems with an observer in the original coordinate. In: IEEE Conference on Decision and Control, pp. 5927–5932 (2013)
4. Bernard, P., Praly, L., Andrieu, V.: Tools for observers based on coordinate augmentation. In: IEEE Conference on Decision and Control (2015)
5. Bernard, P., Praly, L., Andrieu, V.: Expressing an observer in preferred coordinates by transforming an injective immersion into a surjective diffeomorphism. SIAM J. Control Optim. **56**(3), 2327–2352 (2018)
6. Deza, F., Busvelle, E., Gauthier, J., Rakotopara, D.: High gain estimation for nonlinear systems. Syst. Control Lett. **18**, 295–299 (1992)
7. Gauthier, J.P., Hammouri, H., Othman, S.: A simple observer for nonlinear systems application to bioreactors. IEEE Trans. Autom. Control **37**(6), 875–880 (1992)

8. Gibert, V., Burlion, L., Chriette, A., Boada, J., Plestan, F.: Nonlinear observers in vision system: application to civil aircraft landing. In: European Control Conference (2015)
9. Hammouri, H., Ahmed, F., Othman, S.: Observer design based on immersion technics and canonical form. Syst. Control Lett. **114**, 19–26 (2018)
10. Lebastard, V., Aoustin, Y., Plestan, F.: Observer-based control of a walking biped robot without orientation measurement. Robotica **24** (2006)
11. Maggiore, M., Passino, K.: A separation principle for a class of non uniformly completely observable systems. IEEE Trans. Autom. Control **48** (2003)
12. Menini, L., Possieri, C., Tornambe, A.: A "practical" observer for nonlinear systems. In: IEEE Conference on Decision and Control, pp. 3015–3020 (2017)
13. Praly, L., Marconi, L., Isidori, A.: A new observer for an unknown harmonic oscillator. In: Symposium on Mathematical Theory of Networks and Systems (2006)
14. Rapaport, A., Maloum, A.: Design of exponential observers for nonlinear systems by embedding. Int. J. Robust Nonlinear Control **14**, 273–288 (2004)
15. Sadelli, L.: Modélisation, observation et commande de robots vasculaires magnétiques. Ph.D. thesis, Université d'Orléans (2016)
16. Sadelli, L., Fruchard, M., Ferreira, A.: 2D observer-based control of a vascular microrobot. IEEE Trans. Autom. Control **62**(5), 2194–2206 (2017)

Chapter 9
Around Problem 8.1: Augmenting an Injective Immersion into a Diffeomorphism

In [1, 3], we find[1] the following sufficient condition for the augmentation of an immersion into a diffeomorphism.

Lemma 9.1 ([1, 3]) *Let $\mathscr{X}$ be a bounded set, $\mathscr{S}$ be an open subset of $\mathbb{R}^{d_x}$ containing $\mathtt{cl}(\mathscr{X})$, and $T : \mathscr{S} \to T(\mathscr{S}) \subset \mathbb{R}^{d_\xi}$ be an injective immersion. If there exists a bounded open set, $\tilde{\mathscr{S}}$ satisfying*

$$\mathtt{cl}(\mathscr{X}) \subset \tilde{\mathscr{S}} \subset \mathtt{cl}(\tilde{\mathscr{S}}) \subset \mathscr{S}$$

and a C^1 function $\gamma : \mathscr{S} \to \mathbb{R}^{d_\xi \times (d_\xi - d_x)}$ the values of which are $d_\xi \times (d_\xi - d_x)$ matrices satisfying:

$$\det\left(\frac{\partial T}{\partial x}(x) \quad \gamma(x)\right) \neq 0 \qquad \forall x \in \mathtt{cl}(\tilde{\mathscr{S}}) , \tag{9.1}$$

then there exists a strictly positive real number ε such that the following pair[2] $(T_a, \mathscr{S}_a)$ solves Problem 8.1

$$T_a(x, w) \;=\; T(x) \;+\; \gamma(x)\, w \quad , \quad \mathscr{S}_a = \tilde{\mathscr{S}} \times B_\varepsilon(0) \; . \tag{9.2}$$

In other words, an injective immersion T can be augmented into a diffeomorphism T_a if we are able to find $d_\xi - d_x$ columns γ which are C^1 in x and which complement the full-column rank Jacobian $\frac{\partial T}{\partial x}(x)$ into an invertible matrix.

Proof The fact that the Jacobian of T_a is invertible for ε small enough is proved by using (9.1) and the fact that $\tilde{\mathscr{S}}$ is compact in [1]. The injectivity is proved in [3] by compacity of $\mathtt{cl}(\tilde{\mathscr{S}} \times B_{\varepsilon_0}(0))$ and thanks to the implicit function theorem. □

[1]Texts of Chap. 9 are reproduced from [3] with permission from SIAM.

[2]For a positive real number ε and z_0 in $\mathbb{R}^p$, $B_\varepsilon(z_0)$ is the open ball centered at z_0 and with radius ε.

P. Bernard, *Observer Design for Nonlinear Systems*,
Lecture Notes in Control and Information Sciences 479,
https://doi.org/10.1007/978-3-030-11146-5_9

Remark 9.1 Complementing a $d_\xi \times d_x$ full-rank matrix into an invertible one is equivalent to finding $d_\xi - d_x$ independent vectors orthogonal to that matrix. Precisely the existence of γ satisfying (9.1) is equivalent to the existence of a C^1 function $\tilde{\gamma} : \mathtt{cl}(\tilde{\mathscr{S}}) \to \mathbb{R}^{d_\xi \times (d_\xi - d_x)}$ the values of which are full-rank matrices satisfying:

$$\tilde{\gamma}(x)^\top \frac{\partial T}{\partial x}(x) = 0 \qquad \forall x \in \mathtt{cl}(\tilde{\mathscr{S}}) . \tag{9.3}$$

Indeed, $\tilde{\gamma}$ satisfying (9.3) satisfies also (9.1) since the following matrices are invertible

$$\begin{pmatrix} \frac{\partial T}{\partial x}(x)^\top \\ \tilde{\gamma}(x)^\top \end{pmatrix} \begin{pmatrix} \frac{\partial T}{\partial x}(x) & \tilde{\gamma}(x) \end{pmatrix} = \begin{pmatrix} \frac{\partial T}{\partial x}(x)^\top \frac{\partial T}{\partial x}(x) & 0 \\ 0 & \tilde{\gamma}(x)^\top \tilde{\gamma}(x) \end{pmatrix} .$$

Conversely, given γ satisfying (9.1), $\tilde{\gamma}$ defined by the identity below satisfies (9.3) and has full column rank

$$\tilde{\gamma}(x) = \left[I - \frac{\partial T}{\partial x}(x) \left[\frac{\partial T}{\partial x}(x)^\top \frac{\partial T}{\partial x}(x) \right]^{-1} \frac{\partial T}{\partial x}(x)^\top \right] \gamma(x) .$$

9.1 Submersion Case

When $T(\mathtt{cl}(\tilde{\mathscr{S}}))$ is a level set of a submersion, the following complementation result is given in [3].

Theorem 9.1 ([3]) *Let $\mathscr{X}$ be a bounded set, $\tilde{\mathscr{S}}$ be a bounded open set, and $\mathscr{S}$ be an open set satisfying*

$$\mathtt{cl}(\mathscr{X}) \subset \tilde{\mathscr{S}} \subset \mathtt{cl}(\tilde{\mathscr{S}}) \subset \mathscr{S} .$$

Let also $T : \mathscr{S} \to T(\mathscr{S}) \subset \mathbb{R}^{d_\xi}$ be an injective immersion. Assume there exists a C^2 function $F : \mathbb{R}^{d_\xi} \to \mathbb{R}^{d_\xi - d_x}$ which is a submersion[3] *at least on a neighborhood of $T(\tilde{\mathscr{S}})$ satisfying:*

$$F(T(x)) = 0 \qquad \forall x \in \tilde{\mathscr{S}} , \tag{9.4}$$

then, with the C^1 function $x \mapsto \gamma(x) = \frac{\partial F}{\partial \xi}^T(T(x))$, the matrix in (9.1) is invertible for all x in $\tilde{\mathscr{S}}$ and the pair $(T_a, \mathscr{S}_a)$ defined in (9.2) solves Problem 8.1.

Proof For all x in $\mathtt{cl}(\tilde{\mathscr{S}})$, $\frac{\partial T}{\partial x}(x)$ is right invertible and we have $\frac{\partial F}{\partial \xi}(T(x))\frac{\partial T}{\partial x}(x) = 0$. Thus, the rows of $\frac{\partial F}{\partial \xi}(T(x))$ are orthogonal to the column vectors of $\frac{\partial T}{\partial x}(x)$ and are independent since F is a submersion. The Jacobian of T can therefore be completed with $\frac{\partial F}{\partial \xi}^T(T(x))$. The proof is completed with Lemma 9.1. □

[3] $F : \mathbb{R}^{d_\xi} \to \mathbb{R}^n$ with $d_\xi \geq n$ is a submersion on $\mathscr{V}$ if $\frac{\partial F}{\partial \xi}(\xi)$ is full-rank for all ξ in $\mathscr{V}$.

Remark 9.2 Since $\frac{\partial T}{\partial x}$ is of constant rank d_x on $\mathcal{S}$, the existence of such a function F is guaranteed at least locally by the constant rank theorem.

Example 9.1 (*Continuation of Example* 8.1) Elimination of the $\hat{x}_i$ in the four equations given by the injective immersion T defined in (8.4) leads to the function $F(\xi) = \xi_2\xi_3 - \xi_1\xi_4$ satisfying (9.4). It follows that a candidate for complementing:

$$\frac{\partial T}{\partial x}(x) = \begin{pmatrix} 1 & 0 & 0 \\ 0 & 1 & 0 \\ -x_3 & 0 & -x_1 \\ 0 & -x_3 & -x_2 \end{pmatrix} \tag{9.5}$$

is

$$\gamma(x) \;=\; \frac{\partial F}{\partial \xi}(T(x))^\top = (x_2x_3, -x_1x_3, x_2, -x_1)^\top .$$

This vector is nothing but the column of the minors of the matrix (9.5). It gives as determinant $(x_2x_3)^2 + (x_1x_3)^2 + x_2^2 + x_1^2$ which is never zero on $\mathcal{S}$.

Then, it follows from Lemma 9.1, that, for any bounded open set $\tilde{\mathcal{S}}$ such that $\mathcal{X} \subset \mathtt{cl}(\tilde{\mathcal{S}}) \subset \mathcal{S}$ the following function is a diffeomorphism on $\tilde{\mathcal{S}} \times B_\varepsilon(0)$ for ε sufficiently small

$$T_a(x, w) = (x_1 + x_2x_3w, x_2 - x_1x_3w, -x_1x_3 + x_2w, -x_2x_3 - x_1w) .$$

With picking $T_e = T_a$, (8.25) gives us the following observer written in the given x-coordinates augmented with w:

$$\dot{\overbrace{\begin{pmatrix} \hat{x}_1 \\ \hat{x}_3 \\ \hat{x}_2 \\ \hat{w} \end{pmatrix}}} = \begin{pmatrix} 1 & \hat{x}_3\hat{w} & \hat{x}_2\hat{w} & \hat{x}_2\hat{x}_3 \\ -\hat{x}_3\hat{w} & 1 & -\hat{x}_1\hat{w} & -\hat{x}_1\hat{x}_3 \\ -\hat{x}_3 & \hat{w} & -\hat{x}_1 & \hat{x}_2 \\ -\hat{w} & -\hat{x}_3 & -\hat{x}_2 & -\hat{x}_1 \end{pmatrix}^{-1} \left[\begin{pmatrix} \hat{x}_2 - \hat{x}_1\hat{x}_3\hat{w} \\ -\hat{x}_1\hat{x}_3 + \hat{x}_2\hat{w} \\ -\hat{x}_2\hat{x}_3 - \hat{x}_1\hat{w} \\ \mathsf{sat}_{r^3}(\hat{x}_1\hat{x}_3^2) \end{pmatrix} + \begin{pmatrix} Lk_1 \\ L^2k_2 \\ L^3k_3 \\ L^4k_4 \end{pmatrix} [y - \hat{x}_1] \right] \tag{9.6}$$

Unfortunately, the matrix to be inverted is non-singular for $(\hat{x}, \hat{w})$ in $\tilde{\mathcal{S}} \times B_\varepsilon(0)$ only and we have no guarantee that the trajectories of this observer remain in this set. This shows that a further modification transforming T_a into T_e is needed to make sure that $T_e^{-1}(\xi)$ belongs to this set whatever ξ in $\mathbb{R}^4$. This is Problem 8.2. ▲

The drawback of this Jacobian complementation method is that it asks for the knowledge of the function F. It would be better to simply have a universal formula relating the entries of the columns to be added to those of $\frac{\partial T}{\partial x}$.

9.2 The $\tilde{P}[d_\xi, d_x]$ Problem

Finding a universal formula for the Jacobian complementation problem amounts to solving the following problem.

Definition 9.1 ($\tilde{P}[d_\xi, d_x]$ *problem*) For a pair of integers (d_ξ, d_x) such that $0 < d_x < d_\xi$, a C^1 matrix function $\tilde{\gamma} : \mathbb{R}^{m\times n} \to \mathbb{R}^{d_\xi \times (d_\xi - d_x)}$ solves the $\tilde{P}[d_\xi, d_x]$ problem if for any $d_\xi \times d_x$ matrix $\mathfrak{T} = (\mathfrak{T}_{ij})$ of rank d_x, the matrix $\left(\mathfrak{T}\ \tilde{\gamma}(\mathfrak{T})\right)$ is invertible.

As a consequence of a theorem due to Eckmann [6, Sect. 1.7 p. 126] and Lemma 9.1, the following theorem appears in [3].

Theorem 9.2 ([3]) *The $\tilde{P}[d_\xi, d_x]$ problem is solvable by a C^1 function $\tilde{\gamma}$ if and only if the pair (d_ξ, d_x) is in one of the following pairs*

$$(\geq 2, d_\xi - 1) \qquad or \qquad (4, 1) \qquad or \qquad (8, 1). \tag{9.7}$$

Moreover, for each of these pairs and for any bounded set $\mathscr{X}$, any bounded open set $\tilde{\mathscr{S}}$ and any open set $\mathscr{S}$ satisfying

$$\mathtt{cl}(\mathscr{X}) \subset \tilde{\mathscr{S}} \subset \mathtt{cl}(\tilde{\mathscr{S}}) \subset \mathscr{S} \subset \mathbb{R}^{d_x},$$

and any injective immersion $T : \mathscr{S} \to T(\mathscr{S}) \subset \mathbb{R}^{d_\xi}$, the pair $(T_a, \mathscr{S}_a)$ defined in (9.2) with $\gamma(x) = \tilde{\gamma}\left(\frac{\partial T_a}{\partial x}(x)\right)$ solves Problem 8.1.

Proof (“*only if*”) The following theorem is due to Eckmann [6].

Theorem 9.3 (Eckmann’s theorem) *For $d_\xi > d_x$, there exists a continuous function $\tilde{\gamma}_1 : \mathbb{R}^{d_\xi \times d_x} \to \mathbb{R}^{d_\xi}$ with nonzero values and satisfying $\tilde{\gamma}_1(\mathfrak{T})^T \mathfrak{T} = 0$ for any $d_\xi \times d_x$ matrix $\mathfrak{T} = (\mathfrak{T}_{ij})$ of rank d_x if and only if (d_ξ, d_x) is in one of the following pairs*

$$(\geq 2, d_\xi - 1) \qquad or \qquad (even, 1) \qquad or \qquad (7, 2) \qquad or \qquad (8, 3) \tag{9.8}$$

With Remark 9.1, any pair (d_ξ, d_x) for which $\tilde{P}[d_\xi, d_x]$ is solvable must be one in the list (9.3). The pair $(\geq 2, d_\xi - 1)$ is in the list (9.7). For the pair (even, 1), we need to find $d_\xi - 1$ vectors to complement the given one into an invertible matrix. After normalizing the vector $\mathfrak{T}$ so that it belongs to the unit sphere $\mathbb{S}^{d_\xi - 1}$ and projecting each vector $\gamma_i(\mathfrak{T})$ of $\gamma(\mathfrak{T})$ onto the orthogonal complement of $\mathfrak{T}$, this complementation problem is equivalent to asking whether $\mathbb{S}^{d_\xi - 1}$ is parallelizable (since the $\gamma_i(\mathfrak{T})$ will be a basis for the tangent space at $\mathfrak{T}$ for each $\mathfrak{T} \in \mathbb{S}^{d_\xi - 1}$). It turns out that this problems admits solutions only for $d_\xi = 4$ or $d_\xi = 8$ (see [4]). So in the pairs (even, 1) only (4, 1) and (8, 1) are in the list (9.7).

Finally, since $\tilde{P}[1, 6]$ has no solution, the pairs (7, 2) and (8, 3) cannot be in the list (9.7). Indeed, let $\mathfrak{T}$ be a full-column rank $(d_\xi - 1) \times (d_x - 1)$ matrix. $\begin{pmatrix} \mathfrak{T} & 0 \\ 0 & 1 \end{pmatrix}$

is a full-column rank $d_\xi \times d_x$ matrix. If if $\tilde{P}[d_\xi, d_x]$ has a solution, there exist a continuous $(d_\xi - 1) \times (d_\xi - d_x)$ matrix function $\tilde{\gamma}$ and a continuous row vector functions a^T such that $\begin{pmatrix} \tilde{\gamma}(\mathfrak{T}) & \mathfrak{T} & 0 \\ a(\mathfrak{T})^\top & 0 & 1 \end{pmatrix}$ is invertible. This implies that $\begin{pmatrix} \tilde{\gamma}(\mathfrak{T}) & \mathfrak{T} \end{pmatrix}$ is also invertible. So if $\tilde{P}[d_\xi, d_x]$ has a solution, $\tilde{P}[d_\xi - 1, d_x - 1]$ must have one. □

Proof (“*if* ”) For (d_ξ, d_x) equal to (4, 1) or (8, 1) respectively, possible solutions are

$$\tilde{\gamma}(\mathfrak{T}) = \begin{pmatrix} -\mathfrak{T}_2 & \mathfrak{T}_3 & \mathfrak{T}_4 \\ \mathfrak{T}_1 & -\mathfrak{T}_4 & \mathfrak{T}_3 \\ -\mathfrak{T}_4 & -\mathfrak{T}_1 & -\mathfrak{T}_2 \\ \mathfrak{T}_3 & \mathfrak{T}_2 & -\mathfrak{T}_1 \end{pmatrix} \quad , \quad \tilde{\gamma}(\mathfrak{T}) = \begin{pmatrix} \mathfrak{T}_2 & \mathfrak{T}_3 & \mathfrak{T}_4 & \mathfrak{T}_5 & \mathfrak{T}_6 & \mathfrak{T}_7 & \mathfrak{T}_8 \\ -\mathfrak{T}_1 & \mathfrak{T}_4 & -\mathfrak{T}_3 & \mathfrak{T}_6 & -\mathfrak{T}_5 & -\mathfrak{T}_8 & \mathfrak{T}_7 \\ -\mathfrak{T}_4 & -\mathfrak{T}_1 & \mathfrak{T}_2 & \mathfrak{T}_7 & \mathfrak{T}_8 & -\mathfrak{T}_5 & -\mathfrak{T}_6 \\ \mathfrak{T}_3 & -\mathfrak{T}_2 & -\mathfrak{T}_1 & \mathfrak{T}_8 & -\mathfrak{T}_7 & \mathfrak{T}_6 & -\mathfrak{T}_5 \\ -\mathfrak{T}_6 & -\mathfrak{T}_7 & -\mathfrak{T}_8 & -\mathfrak{T}_1 & \mathfrak{T}_2 & \mathfrak{T}_3 & \mathfrak{T}_4 \\ \mathfrak{T}_5 & -\mathfrak{T}_8 & \mathfrak{T}_7 & -\mathfrak{T}_2 & -\mathfrak{T}_1 & -\mathfrak{T}_4 & \mathfrak{T}_3 \\ \mathfrak{T}_8 & \mathfrak{T}_5 & -\mathfrak{T}_6 & -\mathfrak{T}_3 & \mathfrak{T}_4 & -\mathfrak{T}_1 & -\mathfrak{T}_2 \\ -\mathfrak{T}_7 & \mathfrak{T}_6 & \mathfrak{T}_5 & -\mathfrak{T}_4 & -\mathfrak{T}_3 & \mathfrak{T}_2 & -\mathfrak{T}_1 \end{pmatrix}$$

where $\mathfrak{T}_j$ is the jth component of the vector $\mathfrak{T}$. For $d_x = d_\xi - 1$, we have the identity

$$\det\left(\mathfrak{T} \quad \tilde{\gamma}(\mathfrak{T})\right) = \sum_{j=1}^{m} \tilde{\gamma}_j(\mathfrak{T}_{ij})\, M_{j,m}(\mathfrak{T}_{ij})$$

where $\tilde{\gamma}_j$ is the jth component of the vector-valued function $\tilde{\gamma}$ and the $M_{j,m}$, being the cofactors of $(\mathfrak{T}\ \tilde{\gamma}(\mathfrak{T}))$ computed along the last column, are polynomials in the given components $\mathfrak{T}_{ij}$. At least one of the $M_{j,m}$ is nonzero (because they are minors of dimension d_x of $\mathfrak{T}$ which is full-rank). So it is sufficient to take $\tilde{\gamma}_j(\mathfrak{T}_{ij}) = M_{j,m}(\mathfrak{T}_{ij})$. □

In the following example, we show how by exploiting some structure we can reduce the problem to one of these three pairs.

Example 9.2 (*Continuation of Example* 9.1) In Example 9.1, we have complemented the Jacobian (9.5) with the gradient of a submersion and observed that the components of this gradient are actually cofactors. We now know that this is consistent with the case $d_x = d_\xi - 1$. But we can also take advantage from the upper triangularity of the Jacobian (9.5) and complement only the vector $(-x_1, -x_2)$ by, for instance, $(x_2, -x_1)$. The corresponding vector γ is $\gamma(x) = (0, 0, x_2, -x_1)$. Here again, with Lemma 9.1, we know that, for any bounded open set $\tilde{\mathcal{S}}$ such that $\mathtt{cl}(\mathcal{X}) \subset \tilde{\mathcal{S}} \subset \mathtt{cl}(\tilde{\mathcal{S}}) \subset \mathcal{S}$ the function

$$T_a(x, w) \;=\; (x_1\,,\ x_2\,,\ -x_1x_3 + x_2w\,,\ -x_2x_3 - x_1w)$$

is a diffeomorphism on $\tilde{\mathcal{S}} \times B_\varepsilon(0)$. In fact, in this particular case ε can be arbitrary since the Jacobian of T_a is full-rank on $\tilde{\mathcal{S}} \times \mathbb{R}^{d_\xi - d_x}$. With picking $T_e = T_a$, (8.25) gives us the following observer:

$$\dot{\overbrace{\begin{pmatrix}\hat{x}_1\\ \hat{x}_3\\ \hat{x}_2\\ \hat{w}\end{pmatrix}}} = \begin{pmatrix}1 & 0 & 0 & 0\\ 0 & 1 & 0 & 0\\ -\hat{x}_3 & \hat{w} & -\hat{x}_1 & \hat{x}_2\\ -\hat{w} & -\hat{x}_3 & -\hat{x}_2 & -\hat{x}_1\end{pmatrix}^{-1} \left[\begin{pmatrix}\hat{x}_2\\ -\hat{x}_1\hat{x}_3 + \hat{x}_2\hat{w}\\ -\hat{x}_2\hat{x}_3 - \hat{x}_1\hat{w}\\ \mathrm{sat}_{r^3}(\hat{x}_1\hat{x}_3^2)\end{pmatrix} + \begin{pmatrix}Lk_1\\ L^2k_2\\ L^3k_3\\ L^4k_4\end{pmatrix}[y-\hat{x}_1]\right] \tag{9.9}$$

However, the singularity at $\hat{x}_1 = \hat{x}_2 = 0$ remains and Eq. (8.24) is still not satisfied. ▲

Given the very small number of cases where a universal formula exists, we now look for a more general solution to the Jacobian complementation problem.

9.3 Wazewski's Theorem

Historically, the Jacobian complementation problem was first addressed by Wazewski in [7]. His formulation was:

Problem 9.1 (*Wazewski's problem*) Given a continuous function $\mathfrak{r} : \mathcal{S} \subset \mathbb{R}^{d_x} \to \mathbb{R}^{d_\xi \times d_x}$, the values of which are full-rank $d_\xi \times d_x$ matrices look for a continuous function $\gamma : \mathcal{S} \to \mathbb{R}^{d_\xi \times (d_\xi - d_x)}$ such that the matrix $\big(\mathfrak{r}(x)\ \gamma(x)\big)$ is invertible for all x in $\mathcal{S}$.

The difference with the previous section is that here, we look for a continuous function γ of the argument x of $\mathfrak{r}(x)$ instead of continuous functions of $\mathfrak{r}$ itself.

Wazewski established in [7, Theorems 1 and 3] that this other version of the problem admits a far more general solution.

Theorem 9.4 (Wazewski's theorem) *If $\mathcal{S}$, equipped with the subspace topology of $\mathbb{R}^{d_x}$, is a contractible space, then Wazewski's problem admits a solution. Besides, the function γ can be chosen $\mathcal{C}^\infty$ on $\mathcal{S}$.*

Proof The reader is referred to [6, p. 127] or [5, pp. 406–407] and to [7, Theorems 1 and 3] for the complete proof of existence of a continuous function γ when $\mathcal{S}$ is contractible, and to the long version of [3] available in [2] for its modification into a C^1 function thanks to a partition of unity. We rather detail here the constructive main points of the proof originally given by Wazewski in the particular case where $\mathcal{S}$ is a parallelepiped, because it gives an insight on the explicit construction of γ. It is based on Remark 9.1, noting that, if we have the decomposition

$$\mathfrak{r}(x) = \begin{pmatrix}A(x)\\ B(x)\end{pmatrix}$$

with $A(x)$ invertible on some given subset $\mathcal{R}$ of $\mathcal{S}$, then

$$\gamma(x) = \begin{pmatrix}C(x)\\ D(x)\end{pmatrix}$$

makes $\big(\mathfrak{T}(x)\ \gamma(x)\big)$ invertible on $\mathscr{R}$ if and only if $D(x)$ is invertible on $\mathscr{R}$ and we have

$$C(x) = -(A^T(x))^{-1}B(x)^T D(x) \qquad \forall x \in \mathscr{R}\,. \tag{9.10}$$

Thus, C is imposed by the choice of D and choosing D invertible is enough to build γ on $\mathscr{R}$.

Also, if we already have a candidate

$$\begin{pmatrix} A(x) & C_0(x) \\ B(x) & D_0(x) \end{pmatrix}$$

on a boundary $\partial\mathscr{R}$ of $\mathscr{R}$ and $A(x)$ is invertible for all x in $\partial\mathscr{R}$, then, necessarily, $D_0(x)$ is invertible and $C_0(x) = -(A^T(x))^{-1}B(x)^T D_0(x)$ all x in $\partial\mathscr{R}$. Thus, to extend the construction of a continuous function γ inside $\mathscr{R}$ from its knowledge on the boundary $\partial\mathscr{R}$, it suffices to pick D as any invertible matrix satisfying $D = D_0$ on $\partial\mathscr{R}$. Because we can propagate continuously γ from one boundary to the other, Wazewski deduces from these two observations that, it is sufficient to partition the set $\mathscr{S}$ into adjacent sets $\mathscr{R}_i$ where a given $d_\xi \times d_\xi$ minor A_i is invertible. This is possible since $\mathfrak{T}$ is full-rank on $\mathscr{S}$. When $\mathscr{S}$ is a parallelepiped, he shows that there exists an ordering of the $\mathscr{R}_i$ such that the continuity of each D_i can be successively ensured. We illustrate this construction in Example 9.3 below. □

The following corollary is a consequence of Lemma 9.1 and provides another answer to Problem 8.1.

Corollary 9.1 ([3]) *Let $\mathscr{X}$ be a bounded set, $\mathscr{S}$ be an open subset of $\mathbb{R}^{d_x}$ containing $\mathtt{cl}(\mathscr{X})$ and which, equipped with the subspace topology of $\mathbb{R}^{d_x}$, is a contractible space. Let also $T : \mathscr{S} \to T(\mathscr{S}) \subset \mathbb{R}^{d_\xi}$ be an injective immersion. There exists a C^∞ function γ such that, for any bounded open set $\tilde{\mathscr{S}}$ satisfying*

$$\mathtt{cl}(\mathscr{X}) \subset \tilde{\mathscr{S}} \subset \mathtt{cl}(\tilde{\mathscr{S}}) \subset \mathscr{S}$$

we can find a strictly positive real number ε such that the pair $(T_a, \mathscr{S}_a)$ defined in (9.2) solves Problem 8.1.

Example 9.3 Consider the function

$$\mathfrak{T}(x) = \begin{pmatrix} 1 & 0 & 0 \\ 0 & 1 & 0 \\ -x_3 & 0 & -x_1 \\ 0 & -x_3 & -x_2 \\ \frac{\partial \wp}{\partial x_1}x_3 & \frac{\partial \wp}{\partial x_2}x_3 & \wp \end{pmatrix} , \qquad \wp(x_1, x_2) = \max\left\{0, \frac{1}{r^2} - (x_1^2 + x_2^2)\right\}^4 .$$

$\mathfrak{T}(x)$ has full-rank 3 for any x in $\mathbb{R}^3$, since $\wp(x_1, x_2) \neq 0$ when $x_1 = x_2 = 0$. For reasons that will be explained later in Example 11.1, we would like to find two columns, functions of x, which complete this matrix into a square invertible matrix for all x in

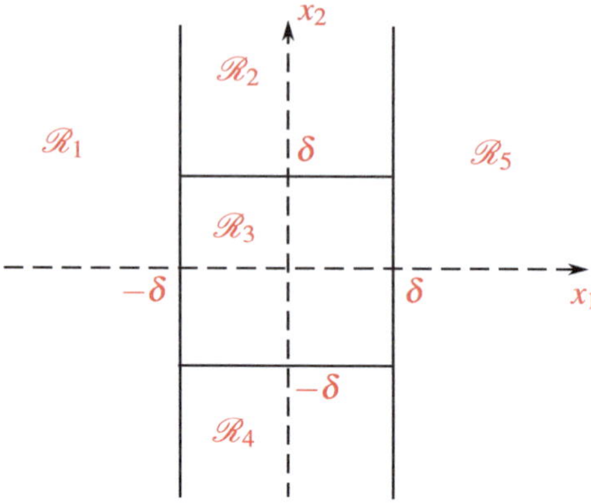

Fig. 9.1 Projections of the regions $\mathscr{R}_i$ on $\mathbb{R}^2$. Figure reproduced from [3] with permission from SIAM

$\mathbb{R}^3$. According to Theorem 9.2, it is not possible to solve $P[3, 5]$, nor $P[1, 3]$ (where we would have taken advantage of the upper triangularity and completed only the vector $(-x_1, -x_2, \wp)^\top$), so that there does not exist a universal completion for those dimensions. We are going to follow Wazewski's instead, since $\mathbb{R}^3$ is contractible.

To follow Wazewski's construction, let δ be a strictly positive real number and consider the following five regions of $\mathbb{R}^3$ (see Fig. 9.1)

$$\mathscr{R}_1 =]-\infty, -\delta] \times \mathbb{R}^2 \quad , \quad \mathscr{R}_2 = [-\delta, \delta] \times [\delta, +\infty] \times \mathbb{R},$$
$$\mathscr{R}_3 = [-\delta, \delta]^2 \times \mathbb{R} \quad , \quad \mathscr{R}_4 = [-\delta, \delta] \times [-\infty, -\delta] \times \mathbb{R} \quad , \quad \mathscr{R}_5 = [\delta, +\infty[\times \mathbb{R}^2.$$

We select δ sufficiently small in such a way that $\wp$ is not 0 in $\mathscr{R}_3$.

We start Wazewski's algorithm in $\mathscr{R}_3$. Here, the invertible minor A is given by rows 1, 2, and 5 of $\mathfrak{T}$ (full-rank lines of $\mathfrak{T}$) and B by rows 3 and 4. With picking D as the identity, C is $(A^T)^{-1}B$ according to (9.10). D gives rows 3 and 4 of γ, and C gives rows 1, 2, and 5 of γ.

Then we move to the region $\mathscr{R}_2$. There the matrix A is given by rows 1, 2, and 4 of $\mathfrak{T}$, B by rows 3 and 5. Also D, along the boundary between $\mathscr{R}_3$ and $\mathscr{R}_2$, is given by rows 3 and 5 of γ obtained in the previous step. We extrapolate this inside $\mathscr{R}_2$ by keeping D constant in planes $x_1 =$ constant. An expression for C and therefore for γ follows.

We do exactly the same thing for $\mathscr{R}_4$.

Then we move to the region $\mathscr{R}_1$. There the matrix A is given by rows 1, 2, and 3 of $\mathfrak{T}$, B by rows 4 and 5. Also D, along the boundary between $\mathscr{R}_1$ and $\mathscr{R}_2$, between $\mathscr{R}_1$ and $\mathscr{R}_3$ and between $\mathscr{R}_1$ and $\mathscr{R}_4$, is given by rows 4 and 5 of γ obtained in the previous steps. We extrapolate this inside $\mathscr{R}_1$ by kipping D constant in planes $x_2 =$ constant. An expression for C and therefore for γ follows.

We do exactly the same thing for $\mathscr{R}_5$.

Note that this construction produces a continuous γ, but we could have extrapolated D in a smoother way to obtain γ as smooth as necessary. ▲

Although Wazewski's method provides a more general answer to the problem of Jacobian complementation than the few solvable $\tilde{P}[d_\xi, d_x]$ problems, the explicit expressions of γ given in Sect. 9.2 are preferred in practice (when the couple (d_ξ, d_x) is in the list (9.7)) to Wazewski's costly computations. In the case where none of those constructions are applicable, we will see in Chap. 11 a universal completion approach, but which necessitates to increase the dimension of T.

We have given several methods to solve Problem 8.1, but to apply Theorem 8.1, we now need to solve Problem 8.2.

References

1. Andrieu, V., Eytard, J.B., Praly, L.: Dynamic extension without inversion for observers. In: IEEE Conference on Decision and Control, pp. 878–883 (2014)
2. Bernard, P., Praly, L., Andrieu, V.: Expressing an observer in given coordinates by augmenting and extending an injective immersion to a surjective diffeomorphism (2018). https://hal.archives-ouvertes.fr/hal-01199791v6
3. Bernard, P., Praly, L., Andrieu, V.: Expressing an observer in preferred coordinates by transforming an injective immersion into a surjective diffeomorphism. SIAM J. Control Optim. **56**(3), 2327–2352 (2018)
4. Bott, R., Milnor, J.: On the parallelizability of the spheres. Bull. Am. Math. Soc. **64**(3), 87–89 (1958)
5. Dugundgi, J.: Topology. Allyn and Bacon, Boston (1966)
6. Eckmann, B.: Mathematical Survey Lectures 1943–2004. Springer, Berlin (2006)
7. Wazewski, T.: Sur les matrices dont les éléments sont des fonctions continues. Compos. Math. **2**, 63–68 (1935)

Chapter 10
Around Problem 8.2: Image Extension of a Diffeomorphism

We study[1] now how a diffeomorphism can be augmented to make its image be the whole set $\mathbb{R}^{d_\xi}$, i.e., to make it surjective, as demanded by Problem 8.2. In certain cases, the construction of the extension is explicit and is illustrated on examples. In particular, we show that solving Problem 8.2 guarantees completeness of solutions of the observer presented in [4] for a bioreactor.

10.1 A Sufficient Condition

There is a rich literature reporting very advanced results on the diffeomorphism extension problem. In the following, some of the techniques are inspired from [5, Chap. 8] and [7, pp. 2, 7–14 and 16–18] (among others). Here we are interested in the particular aspect of this topic which is the diffeomorphism image extension as described by Problem 8.2. A very first necessary condition about this problem is in the following remark.

Remark 10.1 Since T_e, obtained solving Problem 8.2, makes the set $\mathscr{S}$ diffeomorphic to $\mathbb{R}^{d_\xi}$, $\mathscr{S}$ must be contractible.

One of the key technical property which will allow us to solve Problem 8.2 can be phrased as follows.

Definition 10.1 (*Property* $\mathfrak{C}$) An open subset E of $\mathbb{R}^{d_\xi}$ is said to verify *Property* $\mathfrak{C}$ if there exist a C^1 function $\kappa : \mathbb{R}^{d_\xi} \to \mathbb{R}$, a bounded[2] C^1 vector field χ, and a closed set K_0 contained in E such that:

1. $E = \left\{z \in \mathbb{R}^{d_\xi} \ : \ \kappa(z) < 0\right\}$
2. K_0 is globally attractive for χ

---

[1]Texts of Chap. 10 are reproduced from [3] with permission from SIAM.

[2]If not replace χ by $\frac{\chi}{\sqrt{1+|\chi|^2}}$.

P. Bernard, *Observer Design for Nonlinear Systems*,
Lecture Notes in Control and Information Sciences 479,
https://doi.org/10.1007/978-3-030-11146-5_10

3. We have the following transversality property:

$$\frac{\partial \kappa}{\partial z}(z)\chi(z) < 0 \qquad \forall z \in \mathbb{R}^{d_\xi} \,:\, \kappa(z) = 0.$$

The two main ingredients of this condition are the function κ and the vector field χ which, both, have to satisfy the transversality property $\mathfrak{C}.3$. In the case where only the function κ is given satisfying $\mathfrak{C}.1$ and with no critical point on the boundary of E, its gradient could play the role of χ. But then for K_0 to be globally attractive, we need at least to remove all the possible critical points that κ could have outside K_0. This task is performed, for example, on Morse functions in the proof of the h-Cobordism theorem [7]. If however a function χ is given and makes E forward invariant (without necessarily satisfying $\mathfrak{C}.1$ or $\mathfrak{C}.3$), it is proved in [3] that Condition $\mathfrak{C}$ is satisfied by an arbitrarily close superset of E. The main result on the diffeomorphism image extension problem is:

Theorem 10.1 ([3]) *Let $\mathcal{S}_a$ be an open subset of $\mathbb{R}^{d_\xi}$ and T_a: $\mathcal{S}_a \to \mathbb{R}^{d_\xi}$ be a diffeomorphism. If*

(a) either $T_a(\mathcal{S}_a)$ verifies property $\mathfrak{C}$,
(b) or $\mathcal{S}_a$ is C^2-diffeomorphic to $\mathbb{R}^{d_\xi}$ and T_a is C^2,

then for any compact set K in $\mathcal{S}_a$, there exists a diffeomorphism $T_e : \mathcal{S}_a \to \mathbb{R}^{d_\xi}$ solving Problem 8.2.

A sketch of the proof of case (a) of this theorem is given in Sect. 10.2 because it provides an explicit construction of T_e. The main idea of this result (and behind Condition $\mathfrak{C}$) is that the simplest way to build a diffeomorphism is by following the flow of a vector field. The proof of case (b), on the other hand, is not constructive[3] and uses diffeotopy results from [5]. For the time being, we observe that a direct consequence is:

Corollary 10.1 *Let $\mathcal{X}$ be a bounded subset of $\mathbb{R}^{d_x}$, $\mathcal{S}_a$ be an open subset of $\mathbb{R}^{d_\xi}$ containing $K = \mathtt{cl}(\mathcal{X} \times \{0\})$ and $T_a : \mathcal{S}_a \to T_a(\mathcal{S}_a)$ be a diffeomorphism such that*

(a) either $T_a(\mathcal{S}_a)$ verifies property $\mathfrak{C}$,
(b) or $\mathcal{S}_a$ is C^2-diffeomorphic to $\mathbb{R}^{d_\xi}$ and T_a is C^2.

Then, there exists a diffeomorphism $T_e : \mathcal{S}_a \to \mathbb{R}^{d_\xi}$, such that

$$T_e(\mathcal{S}_a) = \mathbb{R}^{d_\xi} \quad , \quad T_e(x,0) \;=\; T_a(x,0) \qquad \forall x \in \mathcal{X} \;.$$

Thus, if besides the pair $(T_a, \mathcal{S}_a)$ solves Problem 8.1, then $(T_e, \mathcal{S}_a)$ solves Problems 8.1 and 8.2.

[3] It is omitted in this book and can be found in [3].

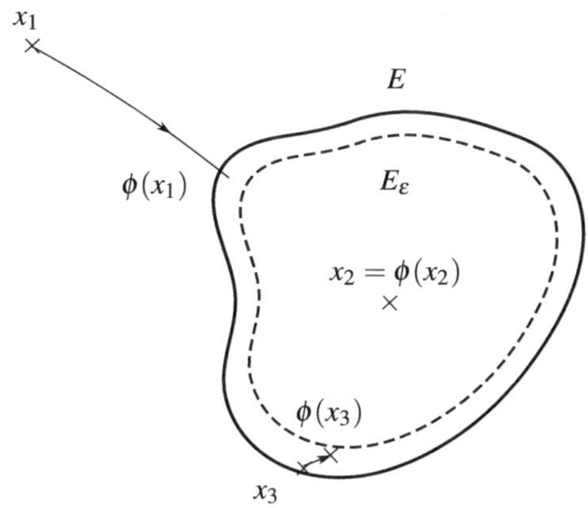

Fig. 10.1 Sketch of the construction of the diffeomorphism ϕ in Lemma 10.1

10.2 Explicit Diffeomorphism Construction for Part (a) of Theorem 10.1

The explicit construction of the diffeomorphism relies on the following important technical lemma.

Lemma 10.1 ([3]) *Let E be an open strict subset of $\mathbb{R}^{d_\xi}$ verifying Condition $\mathfrak{C}$. For any closed subset K of E, lying at a strictly positive distance of the boundary of E, there exists a diffeomorphism ϕ: $\mathbb{R}^{d_\xi} \to E$, such that ϕ is the identity function on K.*

In the long version [2] of [3], a constructive proof of this lemma provides an explicit expression for ϕ which will be used in Example 10.1 and Sect. 10.3. Its construction is illustrated in Fig. 10.1. The rough idea is to follow the flow of the vector field χ given by Condition $\mathfrak{C}$ for a more or less long time depending on the initial point. In fact, denoting E_ε the set where ϕ is the identity, we want to "stuff" in an injective way all the points of $\mathbb{R}^{d_\xi} \setminus E_\varepsilon$ into the "layer" $E \setminus E_\varepsilon$, whose width can be chosen sufficiently small for K to be included in E_ε. To do that, the points of the layer are first "pushed inside" closer to the boundary of E_ε to make room for the points coming from $\mathbb{R}^{d_\xi} \setminus E$. Because this construction is interesting in itself, we give here to the reader a sketch of the proof with the main constructive steps.

Proof Denoting $t \mapsto Z(z,t)$ the (unique) solution to $\dot{z} = \chi(z)$ obtained from Condition $\mathfrak{C}$, going through z at time 0, we start by defining the set of points obtained by following χ from the boundary ∂E of E in a time smaller than $\varepsilon > 0$, namely

$$\Sigma_\varepsilon = \bigcup_{t\in[0,\varepsilon]} Z(\partial E, t) \quad , \quad E_\varepsilon = E \cap (\Sigma_\varepsilon)^{\mathsf{c}} \ .$$

According to Condition $\mathfrak{C}$, χ is bounded and K_0 and K are compact subsets of the open set E. It follows that there exists a strictly positive (maybe infinite) real number ε_∞ such that

$$Z(z,t) \notin K_0 \quad , \quad Z(z,t) \notin K \qquad \forall (z,t) \in \partial E \times [0, 2\varepsilon_\infty[\; .$$

In the following, ε is a real number in $[0, \varepsilon_\infty[$.

The construction of ϕ relies on a function $\mathfrak{t}$ defined on $(E_{2\varepsilon})^{\mathsf{c}}$ by:

$$\kappa\,(Z(z, \mathfrak{t}(z))) = 0 \quad \Longleftrightarrow \quad Z(z, \mathfrak{t}(z)) \in \partial E \; . \tag{10.1}$$

$\mathfrak{t}(z)$ represents the time needed to reach ∂E from z following the vector field χ: It is in $[-2\varepsilon, 0]$ for z in $\Sigma_{2\varepsilon}$ and in $[0, +\infty)$ for z in $\mathbb{R}^{d_\xi} \setminus E$. The assumptions of global attractiveness of the closed set K_0 contained in E open, of transversality of χ to ∂E, and the property of forward-invariance of E, enable to show by the implicit function theorem that this function is well-defined and C^s on $(E_{2\varepsilon})^{\mathsf{c}}$.

It is then possible to establish that $\partial E_{2\varepsilon} \subset \{z \in E \; : \; \mathfrak{t}(z) = -2\varepsilon\}$, and thus extend by continuity the definition of $\mathfrak{t}$ to $\mathbb{R}^{d_\xi}$ by letting

$$\mathfrak{t}(z) \;=\; -2\varepsilon \qquad \forall z \in E_{2\varepsilon} \; .$$

All the properties established for $\Sigma_{2\varepsilon}$ and $E_{2\varepsilon}$ hold also for Σ_ε and E_ε, with in particular

$$\mathfrak{t}(z) \in [-2\varepsilon, -\varepsilon] \qquad \forall z \in E_\varepsilon \setminus E_{2\varepsilon} \; .$$

Let now $\nu : \mathbb{R} \to \mathbb{R}$ be a function such that the function $t \mapsto \nu(t) - t$ is a C^s (decreasing) diffeomorphism from $\mathbb{R}$ onto $]0, +\infty[$ mapping $[-\varepsilon, +\infty[$ onto $]0, \varepsilon]$ and such that

$$\nu(t) - t \;=\; -t \quad \forall t \leq -\varepsilon \; .$$

For instance, a possible candidate is

$$\nu(t) \;=\; \begin{cases} \frac{(t+\varepsilon)^2}{2\varepsilon + t} & t \in [-\varepsilon, +\infty) \\ 0 & \text{otherwise} \; . \end{cases} \tag{10.2}$$

We have

$$\nu(t) > t \qquad \forall t \in \mathbb{R} \quad , \qquad \nu(\mathfrak{t}(z)) \;=\; 0 \qquad \forall z \in E_\varepsilon \setminus E_{2\varepsilon} \; . \tag{10.3}$$

The result is obtained with the function $\phi : \mathbb{R}^{d_\xi} \to E$ by

$$\phi(z) = \begin{cases} Z\,(z, \nu(\mathfrak{t}(z)))\,, & \text{if } z \in (E_{2\varepsilon})^c \; , \\ z, & \text{if } z \in E_{2\varepsilon} \; . \end{cases} \tag{10.4}$$

The image of ϕ is contained in E and like the functions Z, ν, and $\mathfrak{t}$, the function ϕ is C^s on the interior of $(E_{2\varepsilon})^{\mathsf{c}}$. But thanks to (10.3),

$$\phi(z) \;=\; z \qquad \forall z \in E_\varepsilon \; ,$$

so that ϕ is C^s on $(E_{2\varepsilon})^{\mathsf{c}} \cup E_\varepsilon = \mathbb{R}^{d_\xi}$. The rest of the proof consists in proving that ϕ is invertible and its inverse is C^s to conclude that ϕ is a C^s-diffeomorphism from $\mathbb{R}^{d_\xi}$ to E. □

In the case (a) of Theorem 10.1, we suppose that $T_a(\mathscr{S}_a)$ satisfies $\mathfrak{C}$. Now, T_a being a diffeomorphism on an open set $\mathscr{S}_a$, the image of any compact subset K of $\mathscr{S}_a$ is a compact subset of $T_a(\mathscr{S}_a)$. According to Lemma 10.1, there exists a diffeomorphism ϕ from $\mathbb{R}^{d_\xi}$ to $T_a(\mathscr{S}_a)$ which is the identity on $T_a(K)$. Thus, the function $T_e = \phi^{-1} \circ T_a$ solves Problem 8.2 and the theorem is proved.

Example 10.1 (Continuation of Example 9.1) In Example 9.1, we have introduced the function

$$F(\xi) = \xi_2\xi_3 - \xi_1\xi_4 \triangleq \frac{1}{2}\xi^\top M\xi$$

as a submersion on $\mathbb{R}^4 \setminus \{0\}$ satisfying

$$F(T(x)) = 0, \tag{10.5}$$

where T is the injective immersion given in (8.4). With it we have augmented T as

$$T_a(x, w) = T(x) + \frac{\partial F}{\partial \xi}^T (T(x))\, w = T(x) + MT(x)\, w$$

which is a diffeomorphism on $\mathscr{S}_a = \tilde{\mathscr{S}} \times]-\varepsilon, \varepsilon[$ for some strictly positive real number ε.

To modify T_a in T_e satisfying $T_e(\mathscr{S}_a) = \mathbb{R}^4$, we let K be the compact set

$$K = \texttt{cl}(T_a(\mathscr{X} \times \{0\})) \subset T_a(\mathscr{S}_a) \subset \mathbb{R}^4 .$$

With Lemma 10.1, we know that, if $T_a(\mathscr{S}_a)$ verifies property $\mathfrak{C}$, there exists a diffeomorphism ϕ defined on $\mathbb{R}^4$ such that ϕ is the identity function on the compact set K and $\phi(\mathbb{R}^4) = T_e(\mathscr{S}_a)$. In that case, as seen above, the diffeomorphism $T_e = \phi^{-1} \circ T_a$ defined on $\mathscr{S}_a$ is such that $T_e = T_a$ on $\mathscr{X} \times \{0\}$ and $T_e(\mathscr{S}_a) = \mathbb{R}^4$, i.e., would be a solution to Problems 8.1 and 8.2. Unfortunately this is impossible. Indeed, due to the observability singularity at $x_1 = x_2 = 0$, $\tilde{\mathscr{S}}$ (and thus $\mathscr{S}_a$) is not contractible. Therefore, there is no diffeomorphism T_e such that $T_e(\mathscr{S}_a) = \mathbb{R}^4$. We will see in Sect. 11.1 how this problem can be overcome. For the time being, we show that it is still possible to find T_e such that $T_e(\mathscr{S}_a)$ covers “almost all” $\mathbb{R}^4$. The idea is to find an approximation E of $T_a(\mathscr{S}_a)$ verifying property $\mathfrak{C}$ and apply the same method on E. Indeed, if E is close enough to $T_a(\mathscr{S}_a)$, one can expect to have $T_e(\mathscr{S}_a)$ "almost equal to" $\mathbb{R}^4$.

With (10.5) and since $M^2 = I$, we have,

$$F(T_a(x, w)) = |T(x)|^2\, w .$$

Since $\mathscr{S}_a$ is bounded, there exists $\delta > 0$ such that the set

$$E = \left\{\xi \in \mathbb{R}^4 : F(\xi)^2 < \delta\right\}$$

contains $T_a(\mathscr{S}_a)$ and thus the compact set K. Let us show that E verifies property $\mathfrak{C}$.

We pick

$$\kappa(\xi) \;=\; F(\xi)^2 - \delta \;=\; \left(\frac{1}{2}\xi^T M\xi\right)^2 - \delta\,.$$

and consider the vector field

$$\chi(\xi) = -\xi\,.$$

The latter implies the transversality property $\mathfrak{C}$.3 is verified. Besides, the closed set $K_0 = \{0\}$ is contained in E and is globally attractive for the vector field χ.

Then Lemma 10.1 gives the existence of a diffeomorphism $\phi : \mathbb{R}^4 \to E$ which is the identity on K and verifies $\phi(\mathbb{R}^4) = E$. From the sketch of the proof, let E_ε be the set

$$E_\varepsilon = \left\{\xi \in \mathbb{R}^4 : \left(\frac{1}{2}\xi^T M\xi\right)^2 < e^{-4\varepsilon}\,\delta\right\}$$

obtained by following χ from the points in $\partial E = \left\{\xi \in \mathbb{R}^4 : F(\xi)^2 = \delta\right\}$ during a time smaller than ε. It contains K. Let also $\nu : \mathbb{R} \to \mathbb{R}$ defined in (10.2) and $\mathfrak{t} : \mathbb{R}^4 \setminus E_\varepsilon \to \mathbb{R}$ be the functions defined as

$$\mathfrak{t}(\xi) = \frac{1}{4}\ln\frac{\left(\frac{1}{2}\xi^T M\xi\right)^2}{\delta}\,. \tag{10.6}$$

$\mathfrak{t}(\xi)$ is the time that a solution of $\dot{\xi} = \chi(\xi) = -\xi$ with initial condition ξ needs to reach the boundary of E, i.e., $e^{-\mathfrak{t}(\xi)}\xi$ belongs to the boundary of E. From the proof Lemma 10.1, we know the function ϕ: $\mathbb{R}^4 \to E$ defined as:

$$\phi(\xi) = \begin{cases} \xi & , \quad \text{if } \left(\frac{1}{2}\xi^T M\xi\right)^2 \le e^{-4\varepsilon}\delta, \\ e^{-\nu(\mathfrak{t}(\xi))}\xi & , \quad \text{otherwise,} \end{cases} \tag{10.7}$$

is a diffeomorphism $\phi : \mathbb{R}^4 \to E$ which is the identity on K and verifies $\phi(\mathbb{R}^4) = E$.

As explained above, we use ϕ to replace T_a by the diffeomorphism $T_e = \phi^{-1} \circ T_a$ also defined on $\mathscr{S}_a$. But, because $T_a(\mathscr{S}_a)$ is a strict subset of E, $T_e(\mathscr{S}_a)$ is a strict subset of $\mathbb{R}^4$, i.e., Eq. (8.24) is not satisfied. Nevertheless, for any trajectory of the observer $t \mapsto \hat{\xi}(t)$ in $\mathbb{R}^4$, our estimate defined by $(\hat{x}, \hat{w}) = T_e{}^{-1}(\hat{\xi})$ will be such that $T_a(\hat{x}, \hat{w})$ remains in E, along this trajectory, i.e., $|T(\hat{x})|^2\,\hat{w} < \delta$. This ensures that, far from the observability singularity where $|T(\hat{x})| = 0$, $\hat{w}$ remains sufficiently small to keep the invertibility of the Jacobian of T_e. But we still have a problem close to the observability singularity, i.e., when $(\hat{x}_1, \hat{x}_2)$ is close to the origin. We shall see in Sect. 11.1 how to avoid this difficulty via a better choice of the initial injective immersion T. ▲

10.3 Application: Bioreactor

As a less academic illustration, let us consider the bioreactor model used in [4]

$$\left\{\begin{aligned} \dot{x}_1 &= \frac{a_1 x_1 x_2}{a_2 x_1 + x_2} - u x_1 \\ \dot{x}_2 &= -\frac{a_3 a_1 x_1 x_2}{a_2 x_1 + x_2} - u x_2 + u a_4 \end{aligned}\right. \quad , \quad y = x_1 \tag{10.8}$$

where the a_i's are strictly positive real numbers and the control u verifies $0 < u_{min} < u(t) < u_{max} < a_1$. This system evolves in the set

$$\mathcal{S} = \left\{ x \in \mathbb{R}^2 \, : \, x_1 > 0 \, , \, x_2 > -a_2 x_1 \right\}$$

which is forward invariant. The autonomous part of this model has a similar structure to the other bioreactor model (8.10) that we studied in Chap. 8, but with a map

$$\mu(x_1, x_2) = \frac{a_1 x_2}{a_2 x_1 + x_2}$$

such that $x_2 \mapsto \mu(x_1, x_2)$ is monotonic on $(-a_2 x_1, +\infty)$ for any $x_1 > 0$. Therefore, this time, the map $T : \mathcal{S} \to \mathbb{R}^2$ defined as:

$$T(x_1, x_2) = (x_1, \dot{x}_1|_{u=0}) = \left(x_1, \frac{a_1 x_1 x_2}{a_2 x_1 + x_2} \right) .$$

is a diffeomorphism on $\mathcal{S}$ onto

$$T(\mathcal{S}) = \left\{ \xi \in \mathbb{R}^2 \, : \, \xi_1 > 0 \, , \, a_1 \xi_1 > \xi_2 \right\} .$$

This means that the drift system is strongly differentially observable[4] of order $d_x = 2$. Actually, the full bioreactor dynamics (10.8) are strongly differentially observable[5] of order $d_x = 2$ for any input, and therefore, uniformly instantaneously observable. According to Theorem 7.2, the image by T of (10.8) is a Lipschitz triangular form

$$\left\{\begin{aligned} \dot{\xi}_1 &= \xi_2 + \mathfrak{g}_1(\xi_1) u \\ \dot{\xi}_2 &= \varphi_2(\xi_1, \xi_2) + \mathfrak{g}_2(\xi_1, \xi_2) u \end{aligned}\right.$$

[4]See Definition 5.3.

[5]See Definition 5.2.

for which the following high-gain observer can be built

$$\begin{cases} \dot{\xi}_1 = \xi_2 + \mathfrak{g}_1(\xi_1)u - Lk_1(\xi_1 - y) \\ \dot{\xi}_2 = \varphi_2(\xi_1, \xi_2) + \mathfrak{g}_2(\xi_1, \xi_2)u - L^2k_2(\xi_1 - y) \end{cases} \tag{10.9}$$

where k_1 and k_2 are positive real numbers and L sufficiently large.

As observed in [4], T being a diffeomorphism, the dynamics of this observer in the x-coordinates are

$$\dot{\hat{x}} = \begin{pmatrix} \frac{a_1\hat{x}_1\hat{x}_2}{a_2\hat{x}_1+\hat{x}_2} - u\hat{x}_1 \\ -\frac{a_3a_1\hat{x}_1\hat{x}_2}{a_2\hat{x}_1+\hat{x}_2} - u\hat{x}_2 + ua_4 \end{pmatrix} + \begin{pmatrix} 1 & 0 \\ -1 & \frac{(a_2\hat{x}_1+\hat{x}_2)^2}{a_1a_2\hat{x}_1^2} \end{pmatrix} \begin{pmatrix} Lk_1 \\ L^2k_2 \end{pmatrix} (\xi_1 - y) \, . \tag{10.10}$$

Unfortunately the right-hand side is singular at $\hat{x}_1 = 0$ or $\hat{x}_2 = -a_1\hat{x}_1$. $\mathcal{S}$ being forward invariant, the system trajectories stay away from the singularities, but nothing guarantees the same property holds for the observer trajectories given by (10.10). In other words, since T is already a diffeomorphism, Problem 8.1 is solved with $d_\xi = d_x$, $T_a = T$ and $\mathcal{S}_a = \mathcal{S}$, but (8.24) is not satisfied, i.e., $T_a(\mathcal{S}) \neq \mathbb{R}^2$. This means that we must extend the image of $T_a = T$ to make it surjective, namely Problem 8.2 must be solved.

To construct the extension T_e of T_a, we view the image $T_a(\mathcal{S}_a)$ as the intersection $T_a(\mathcal{S}_a) = E_1 \cap E_2$ with:

$$E_1 = \left\{(\xi_1, \xi_2) \in \mathbb{R}^2, \ \xi_1 > \varepsilon_1\right\} \quad , \qquad E_2 = \left\{(\xi_1, \xi_2) \in \mathbb{R}^2, \ a_1\xi_1 > \xi_2\right\} \, .$$

This exhibits the fact that $T_a(\mathcal{S}_a)$ does not satisfy the property $\mathfrak{C}$ since its boundary is not C^1. We could smoothen this boundary to remove its "corner." But we prefer to exploit its particular "shape" and proceed as follows:

1. We build a diffeomorphism $\phi_1 \colon \mathbb{R}^2 \to E_1$ which acts on ξ_1 without changing ξ_2.
2. We build a diffeomorphism $\phi_2 \colon \mathbb{R}^2 \to E_2$ which acts on ξ_2 without changing ξ_1.
3. Denoting $\phi = \phi_2 \circ \phi_1 : \mathbb{R}^2 \to E_1 \cap E_2$, we take $T_e = \phi^{-1} \circ T_a : \mathcal{S}_a \to \mathbb{R}^2$.

To build ϕ_1 and ϕ_2, we follow the procedure given in the proof of Lemma 10.1 since E_1 and E_2 satisfy property $\mathfrak{C}$ with:

$$\kappa_1(\xi) = \varepsilon_1 - \xi_1 \, , \ \kappa_2(\xi) = \xi_2 - a_1\xi_1$$

$$\chi_1(\xi) = \begin{pmatrix} -(\xi_1 - 1) \\ 0 \end{pmatrix} \, , \ \chi_2(\xi) = \begin{pmatrix} 0 \\ -(\xi_2 + 1) \end{pmatrix} .$$

By following the same steps as in Example 10.1, with ε an arbitrary small strictly positive real number and ν defined in (10.6), we obtain:

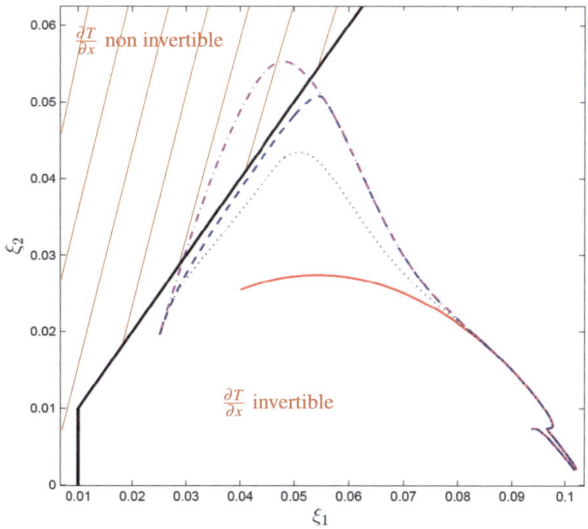

Fig. 10.2 Bioreactor and observers solutions in the ξ-coordinates: singularity locus (solid black), plant's trajectory (solid red), observer trajectory (dashed magenta), its image by ϕ (dashed blue), observer trajectory with constraint term (dotted black). Figure reproduced from [3] with permission from SIAM

$$
\left|\begin{array}{l}
\mathfrak{t}_1(\xi) = \ln \frac{1-\xi_1}{1-\varepsilon} \\
E_{\varepsilon,1} = \left\{(\xi_1, \xi_2) \in \mathbb{R}^2,\ \xi_1 > 1 - \frac{1-\varepsilon}{e^{\varepsilon}}\right\} \\
\phi_1(\xi) = \begin{cases} \xi & , \text{ if } \xi \in E_{\varepsilon,1} \\ \frac{\xi_1 - 1}{e^{\nu(\mathfrak{t}_1(\xi))}} + 1 , & \text{otherwise} \end{cases}
\end{array}\right.
\qquad
\left|\begin{array}{l}
\mathfrak{t}_2(\xi) = \ln \frac{\xi_2+1}{a_1\xi_1+1} , \\
E_{\varepsilon,2} = \left\{(\xi_1, \xi_2) \in \mathbb{R}^2,\ \xi_2 \leq \frac{a_1\xi_1+1}{e^{\varepsilon}} - 1\right\} \\
\phi_2(\xi) = \begin{cases} \xi & , \text{ if } \xi \in E_{\varepsilon,2} \\ \frac{\xi_2 + 1}{e^{\nu(\mathfrak{t}_2(\xi))}} - 1 , & \text{otherwise} \end{cases}
\end{array}\right.
\tag{10.11}
$$

We remind the reader that, in the ξ-coordinates, the observer dynamics are not modified. The difference between using T or T_e is seen in the $\hat{x}$-coordinates only. And, by construction it has no effect on the system trajectories since we have

$$T(x) = T_e(x) \quad \forall x \in \mathscr{S} \text{ “} - \varepsilon\text{”} .$$

As a consequence the difference between T and T_e is significant only during the transient, making sure, for the latter, that $\hat{x}$ never reaches a singularity of the Jacobian of T_e.

We present in Fig. 10.2 the results in the ξ coordinates (to allow us to see the effects of both T and T_e) of a simulation with (similar to [4]):

$$
\begin{gathered}
a_1 = a_2 = a_3 = 1 \ , \ a_4 = 0.1 \\
u(t) = 0.08 \text{ for } t \leq 10 \ , \quad = 0.02 \text{ for } 10 \leq t \leq 20 \ , \quad = 0.08 \text{ for } t \geq 20 \\
x(0) = (0.04, 0.07), \quad \hat{x}(0) = (0.03, 0.09), \quad L = 5.
\end{gathered}
$$

The solid black curves are the singularity locus. The red curve represents the bioreactor solution. The magenta curve represents the solution of the observer built with T_e. It evolves freely in $\mathbb{R}^2$ according to the dynamics (10.9), not worried by any constraints. The blue curve represents its image by ϕ which brings it back inside the constrained domain where T^{-1} can then be used. This means these two curves represent the same object but viewed in different coordinates.

The solution of the observer built with T would coincide with the magenta curve up to the point it reaches one solid black curve of a singularity locus. At that point it leaves $T(\mathscr{S})$ and consequently stops existing in the x-coordinates.

As proposed in [1, 6], instead of keeping the raw dynamics (10.9) untouched as above, another solution would be to modify them to force ξ to remain in the set $T(\mathscr{S})$. For instance, taking advantage of the convexity of this set, the modification proposed in [1] consists in adding to (10.9) the term

$$\mathscr{M}(\xi) = -\ell\, P^{-1}\, \frac{\partial \mathfrak{h}}{\partial \xi}(\xi)^T\, \mathfrak{h}(\xi) \quad , \qquad \mathfrak{h}(\xi) \;=\; \begin{pmatrix} \max\{\kappa_1(\xi)+\varepsilon, 0\}^2 \\ \max\{\kappa_2(\xi)+\varepsilon, 0\}^2 \end{pmatrix} \tag{10.12}$$

with P a symmetric positive definite matrix depending on (k_1, k_2, L), ε an arbitrary small real number and ℓ a sufficiently large real number. The solution corresponding to this modified observer dynamics is shown in Fig. 10.2 with the dotted black curve. As expected it stays away from the singularities locus in a very efficient way. But, for this method to apply, we have the restriction that $T(\mathscr{S})$ should be convex, instead of satisfying the less restrictive property $\mathfrak{C}$. Moreover, to guarantee that $\hat{\xi}$ stays in $T(\mathscr{S})$, ℓ has to be large enough and even larger when the measurement noise is larger. Indeed, the role of the correction term $\mathscr{M}$ is to compensate for the observer dynamics around the singularity locus in order to force $\hat{\xi}$ to stay in the *safe set*. On the contrary, when the observer is built with T_e, there is no need to tune any parameter properly to obtain convergence, at least theoretically. Nevertheless, there may be some numerical problems when ξ becomes too large or equivalently $\phi(\xi)$ is too close to the boundary of $T(\mathscr{S})$. To overcome this difficulty, we can select the "thickness" of the layer (parameter ε in (10.11)) sufficiently large. Actually instead of "opposing" the two methods, they can be combined when possible. The modification (10.12) makes sure that ξ does not go too far outside the domain, and T_e makes sure that $\hat{x}$ does not cross the singularity locus.

10.4 Conclusion

Joining Corollaries 9.1 and 10.1, we obtain the following answer to our problem:

Corollary 10.2 *Let $\mathscr{X}$ be a bounded subset of $\mathbb{R}^{d_x}$, $\mathscr{S}$ be an open subset of $\mathbb{R}^{d_x}$ and $T : \mathscr{S} \to \mathbb{R}^{d_\xi}$ be an injective immersion. Assume there exists an open-bounded contractible set $\tilde{\mathscr{S}}$ which is C^2-diffeomorphic to $\mathbb{R}^{d_x}$ and such that*

$$\mathtt{cl}(\mathscr{X}) \subset \tilde{\mathscr{S}} \subset \mathtt{cl}(\tilde{\mathscr{S}}) \subset \mathscr{S} \,.$$

There exists a strictly positive number ε *and a diffeomorphism* $T_e : \mathcal{S}_a \to \mathbb{R}^{d_\xi}$ *with* $\mathcal{S}_a = \tilde{\mathcal{S}} \times B_\varepsilon(0)$*, such that*

$$T_e(x, 0) = T(x) \quad \forall x \in \mathcal{X} \quad , \qquad T_e(\mathcal{S}_a) = \mathbb{R}^{d_\xi} \ ,$$

namely $(T_e, \mathcal{S}_a)$ *solves Problems 8.1–8.2.*

We conclude that if $\mathcal{X}$, $\mathcal{S}$, and T given by Assumption 8.1 verify the conditions of Corollary 10.2, then Problems 8.1–8.2 can be solved and Theorem 8.1 holds, i.e., an observer can be expressed in the given x-coordinates.

References

1. Astolfi, D., Praly, L.: Output feedback stabilization for SISO nonlinear systems with an observer in the original coordinate. In: IEEE Conference on Decision and Control, pp. 5927–5932 (2013)
2. Bernard, P., Praly, L., Andrieu, V.: Expressing an observer in given coordinates by augmenting and extending an injective immersion to a surjective diffeomorphism (2018). https://hal.archives-ouvertes.fr/hal-01199791v6
3. Bernard, P., Praly, L., Andrieu, V.: Expressing an observer in preferred coordinates by transforming an injective immersion into a surjective diffeomorphism. SIAM J. Control Optim. **56**(3), 2327–2352 (2018)
4. Gauthier, J.P., Hammouri, H., Othman, S.: A simple observer for nonlinear systems application to bioreactors. IEEE Trans. Autom. Control **37**(6), 875–880 (1992)
5. Hirsch, M.: Differential Topology. Springer, Berlin (1976)
6. Maggiore, M., Passino, K.: A separation principle for a class of non uniformly completely observable systems. IEEE Trans. Autom. Control **48** (2003)
7. Milnor, J.: Lectures on the h-Cobordism Theorem. Notes by L. Siebenmann and J. Sondow. Princeton University Press, Princeton (1965)

Chapter 11
Generalizations and Examples

Throughout Chaps. 9 and 10, we have seen (in particular in Corollary 10.2) conditions allowing to solve Problems 8.1 and 8.2 when Assumption 8.1 holds, $\mathcal{X}$ is bounded, and $\mathcal{S}$ is contractible. However, it can happen that those conditions are not satisfied and we will see in this chapter how to solve both Problems 8.1 and 8.2 via a better choice of the data given by Assumption 8.1, namely T and $\varphi_{\mathcal{T}}$. In particular, this enables to write an observer in the x-coordinates for the oscillator with unknown frequency (8.2) both via the high gain (8.5) and Luenberger (8.7) designs.

Finally, we will see, through an application in aircraft landing, how the methodology presented in this Part III can be extended to the case where the transformation T is time-varying.

11.1 Modifying T and $\varphi_{\mathcal{T}}$ given by Assumption 8.1

The sufficient conditions[1] given in Chaps. 9 and 10, to solve Problems 8.1 and 8.2 in order to fulfill the requirements of Theorem 8.1, impose conditions on the dimensions or on the domain of injectivity $\mathcal{S}$ which are not always satisfied: contractibility for Jacobian complementation and diffeomorphism extension, limited number of pairs (d_ξ, d_x) for the $\tilde{P}[d_\xi, d_x]$ problem, etc. Expressed in terms of our initial problem, these conditions are limitations on the data f, h, and T that we have considered. In the following, we show by means of examples that, in some cases, these data can be modified in such a way that the various tools apply and give a satisfactory solution. Such modifications are possible since we restrict our attention to system solutions which remain in $\mathcal{X}$. Therefore, the data f, h, and T can be arbitrarily modified outside this set. For example, "fictitious" components can be added to the measured output y as long as their value is known on $\mathcal{X}$.

[1]The text of Sect. 11.1 is reproduced from [2] with permission from SIAM.

P. Bernard, *Observer Design for Nonlinear Systems*,
Lecture Notes in Control and Information Sciences 479,
https://doi.org/10.1007/978-3-030-11146-5_11

11.1.1 For Contractibility

It may happen that the set $\mathcal{S}$ attached to T is not contractible, for example due to an observability singularity. We have seen that Jacobian complementation and image extension may be prevented by this (see Theorem 9.4 and Remark 10.1). A possible approach to overcome this difficulty when we know the system trajectories stay away from the singularities is to add a fictitious output traducing this information.

Example 11.1 (*Continuation of Example* 9.2) The observer (9.9) we have obtained at the end of Example 9.2 for the harmonic oscillator with unknown frequency is not satisfactory because of the singularity at $\hat{x}_1 = \hat{x}_2 = 0$. To overcome this difficulty we add, to the given measurement $y = x_1$, the following one

$$y_2 = h_2(x) \;=\; \wp(x_1, x_2)\, x_3$$

with

$$\wp(x_1, x_2) \;=\; \max\left\{0, \frac{1}{r^2} - (x_1^2 + x_2^2)\right\} . \tag{11.1}$$

By construction, this function is zero on $\mathcal{X}$ and y_2 can thus be considered as an extra measurement with zero as constant value. The interest of y_2 is to give access to x_3 even at the singularity $x_1 = x_2 = 0$. Indeed, consider the new function T defined as

$$T(x) \;=\; (x_1\,,\ x_2\,,\ -x_1x_3\,,\ -x_2x_3\,,\ \wp(x_1, x_2)\, x_3) \ . \tag{11.2}$$

T is C^1 on $\mathbb{R}^3$ and its Jacobian is

$$\frac{\partial T}{\partial x}(x) \;=\; \begin{pmatrix} 1 & 0 & 0 \\ 0 & 1 & 0 \\ -x_3 & 0 & -x_1 \\ 0 & -x_3 & -x_2 \\ \frac{\partial \wp}{\partial x_1} x_3 & \frac{\partial \wp}{\partial x_2} x_3 & \wp \end{pmatrix} , \tag{11.3}$$

which has full-rank 3 on $\mathbb{R}^3$, since $\wp(x_1, x_2) \neq 0$ when $x_1 = x_2 = 0$. It follows that the singularity has disappeared and this new T is an injective immersion on the entire $\mathbb{R}^3$ which is contractible.

We have shown in Example 9.3 how Wazewski's algorithm allows us to get in this case a C^2 function $\gamma : \mathbb{R}^3 \to \mathbb{R}^4$ satisfying:

$$\det\left(\frac{\partial T}{\partial x}(x)\, \gamma(x)\right) \;\neq\; 0 \qquad \forall x \in \mathbb{R}^3 \ .$$

This gives us $T_a(x, w) \;=\; T(x) + \gamma(x)w$ which is a C^2-diffeomorphism on $\mathbb{R}^3 \times B_\varepsilon(0)$, with ε sufficiently small. Furthermore, $\mathcal{S}_a = \mathbb{R}^3 \times B_\varepsilon(0)$ being now diffeomorphic to $\mathbb{R}^5$, Corollary 10.1 applies and provides an extension T_e of T_a satisfying Problems 8.1 and 8.2. ▲

11.1.2 For a Solvable $\tilde{P}[d_\xi, d_x]$ Problem

If we are in a case that cannot be reduced to a solvable $\tilde{P}[d_\xi, d_x]$ problem, we may try to modify d_ξ by adding arbitrary rows to $\frac{\partial T}{\partial x}$. This technique is illustrated with the following example.

Example 11.2 (*Continuation of Example* 11.1) In Example 11.1, by adding the fictitious measured output $y_2 = h_2(x)$, we have obtained another function T for the harmonic oscillator with unknown frequency which is an injective immersion on $\mathbb{R}^3$. In this case, we have $d_x = 3$ and $d_\xi = 5$ which gives a pair not in (9.7). But, as already exploited in Example 9.2, the first two rows of the Jacobian $\frac{\partial T}{\partial x}$ in (11.3) are independent for all x in $\mathbb{R}^3$. It follows that our Jacobian complementation problem reduces to complement the vector $(-x_1, -x_2, \wp(x_1, x_2))^\top$. This is a problem with pair (3, 1) which is still not in the list (9.7). Instead, the pair (4, 1) is, so that the vector $(-x_1, -x_2, \wp(x_1, x_2), 0)^\top$ can be complemented via a universal formula. We have added a zero component, without changing the full-rank property. Actually this vector is extracted from the Jacobian of

$$T(x) = (x_1\,,\ x_2\,,\ -x_1x_3\,,\ -x_2x_3\,,\ \wp(x_1, x_2)\,x_3\,,\ 0)\ . \tag{11.4}$$

In the high-gain observer paradigm, this added zero can come from another (fictitious) measured output $y_3 = 0$. As we saw in the proof of Theorem 9.2, a complement of $(-x_1, -x_2, \wp(x_1, x_2), 0)^\top$ is

$$\begin{pmatrix} x_2 & -\wp(x_1, x_2) & 0 \\ -x_1 & 0 & -\wp(x_1, x_2) \\ 0 & -x_1 & -x_2 \\ \wp(x_1, x_2) & x_2 & -x_1 \end{pmatrix}$$

and thus a complement of $\frac{\partial T}{\partial x}(x)$ is

$$\gamma(x) = \begin{pmatrix} 0 & 0 & 0 \\ 0 & 0 & 0 \\ x_2 & -\wp(x_1, x_2) & 0 \\ -x_1 & 0 & -\wp(x_1, x_2) \\ 0 & -x_1 & -x_2 \\ \wp(x_1, x_2) & x_2 & -x_1 \end{pmatrix}$$

which gives with (9.2)

$$T_a(x,w) = \Big(x_1 \,,\, x_2 \,,\, [-x_1x_3 + x_2w_1 - \wp(x_1,x_2)w_2] \,,\, [-x_2x_3 - x_1w_1 - \wp(x_1,x_2)w_3] \,,\\ [\wp(x_1,x_2)x_3 - x_1w_2 - x_2w_3] \,,\, [\wp(x_1,x_2)w_1 + x_2w_2 - x_1w_3)]\Big) .$$

The determinant of the Jacobian of T_a thus defined is $(x_1^2 + x_2^2 + \wp(x_1,x_2)^2)^2$ which is nowhere 0 on $\mathbb{R}^6$. Hence, T_a is locally invertible. Actually, it is a diffeomorphism from $\mathbb{R}^6$ onto $\mathbb{R}^6$ since we can express $\xi = T_a(x,w)$ as

$$\begin{pmatrix} x_1 \\ x_2 \end{pmatrix} = \begin{pmatrix} \xi_1 \\ \xi_2 \end{pmatrix}$$

$$\begin{pmatrix} -\xi_1 & \xi_2 & -\wp(\xi_1,\xi_2) & 0 \\ -\xi_2 & -\xi_1 & 0 & -\wp(\xi_1,\xi_2) \\ \wp(\xi_1,\xi_2) & 0 & -\xi_1 & -\xi_2 \\ 0 & \wp(\xi_1,\xi_2) & \xi_2 & -\xi_1 \end{pmatrix} \begin{pmatrix} x_3 \\ w_1 \\ w_2 \\ w_3 \end{pmatrix} = \begin{pmatrix} \xi_3 \\ \xi_4 \\ \xi_5 \\ \xi_6 \end{pmatrix} ,$$

where the matrix on the left is invertible by construction. Since $T_a(\mathbb{R}^6) = \mathbb{R}^6$, there is no need for an image extension and we simply take $T_e = T_a$. To have all the assumptions of Theorem 8.1 satisfied, it remains to find a function φ such that $(\mathscr{T}_{ex}, \varphi)$ is in the set $\varphi\mathscr{T}$, the function $\mathscr{T}_{ex}$ being the x-component of the inverse of T_e. Since the first four components of T are the same as in (8.4), the first four components of φ are given in (8.6). It remains to define the dynamics of $\hat{\xi}_5$ and $\hat{\xi}_6$. Exploiting the fact that, for x in $\mathscr{X}$,

$$y_2 = 0 \quad , \quad \dot{y}_2 = \dot{\overbrace{\wp(x_1,x_2)x_3}} = 0 \quad , \qquad y_3 = 0 \quad , \quad \dot{y}_3 = 0 \, ,$$

one can simply choose

$$\dot{\hat{\xi}}_5 = 0 - a(\hat{\xi}_5 - y_2) = -a\,\hat{\xi}_5 \quad , \quad \dot{\hat{\xi}}_6 = 0 - b(\hat{\xi}_6 - y_3) = -b\,\hat{\xi}_6$$

for some strictly positive numbers a and b. This finally leads to the observer dynamics

$$\varphi(\xi, \hat{x}, y) \;=\; \begin{pmatrix} \xi_2 + Lk_1(y - \hat{x}_1) \\ \xi_3 + L^2k_2(y - \hat{x}_1) \\ \xi_4 + L^3k_3(y - \hat{x}_1) \\ \mathrm{sat}_{r^3}(\hat{x}_1\hat{x}_3^2) + L^4k_4(y - \hat{x}_1) \\ -a\,\xi_5 \\ -b\,\xi_6 \end{pmatrix} .$$

With picking L large enough, φ can be paired with any function $\mathscr{T} : \mathbb{R}^6 \to \mathbb{R}^6$ which is locally Lipschitz, and thus in particular with $\mathscr{T}_{ex}$. Therefore, Theorem 8.1 applies and gives the following observer for the harmonic oscillator with unknown frequency

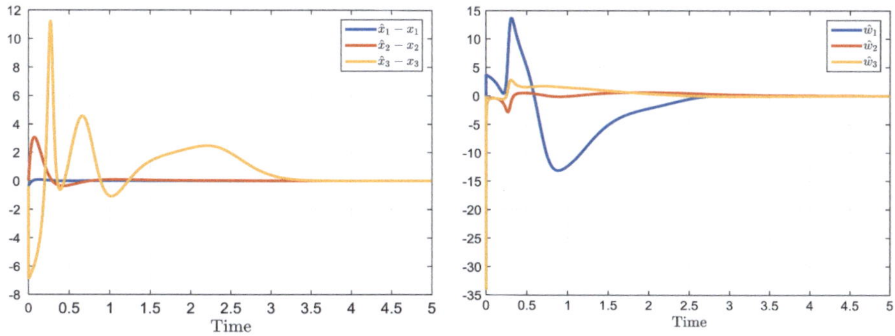

Fig. 11.1 High-gain observer (11.5) initialized at $\hat{x}_1 = \hat{x}_2 = \hat{x}_3 = 0$ at the singularity, $L = 3, k_1 = 10$, $k_2 = 35$, $k_3 = 50$, $k_4 = 24$ for the oscillator (8.2) initialized at (0.35, 0, 0.4). The simulation was done with a variable step Euler algorithm

$$\begin{pmatrix} \dot{\hat{x}}_1 \\ \dot{\hat{x}}_2 \\ \dot{\hat{x}}_3 \\ \dot{\hat{w}}_1 \\ \dot{\hat{w}}_2 \\ \dot{\hat{w}}_3 \end{pmatrix} = \begin{pmatrix} 1 & 0 & 0 & 0 & 0 & 0 \\ 0 & 1 & 0 & 0 & 0 & 0 \\ -\hat{x}_3 - \frac{\partial \wp}{\partial \hat{x}_1}\hat{w}_2 & \hat{w}_1 - \frac{\partial \wp}{\partial x_2}\hat{w}_2 & -\hat{x}_1 & \hat{x}_2 & -\wp & 0 \\ -\hat{w}_1 - \frac{\partial \wp}{\partial \hat{x}_1}\hat{w}_3 & -\hat{x}_3 - \frac{\partial \wp}{\partial x_2}\hat{w}_3 & -\hat{x}_2 & -\hat{x}_1 & 0 & -\wp \\ \frac{\partial \wp}{\partial x_1}\hat{x}_3 - \hat{w}_2 & \frac{\partial \wp}{\partial x_2}\hat{x}_3 - \hat{w}_3 & \wp & 0 & -\hat{x}_1 & -\hat{x}_2 \\ \frac{\partial \wp}{\partial x_1}\hat{w}_1 - \hat{w}_3 & \frac{\partial \wp}{\partial x_2}\hat{w}_1 + \hat{w}_2 & 0 & \wp & \hat{x}_2 & -\hat{x}_1 \end{pmatrix}^{-1} \times \tag{11.5}$$

$$\times \begin{pmatrix} \hat{x}_2 + Lk_1(y - \hat{x}_1) \\ [-\hat{x}_1\hat{x}_3 + \hat{x}_2\hat{w}_1 - \wp(\hat{x}_1, \hat{x}_2)\hat{w}_2] + L^2 k_2(y - \hat{x}_1) \\ [-\hat{x}_2\hat{x}_3 - \hat{x}_1\hat{w}_1 - \wp(\hat{x}_1, \hat{x}_2)\hat{w}_3] + L^3 k_3(y - \hat{x}_1) \\ \mathsf{sat}_{r^3}(\hat{x}_1\hat{x}_3^2) + L^4 k_4(y - \hat{x}_1) \\ -a\,[\wp(\hat{x}_1, \hat{x}_2)\hat{x}_3 - \hat{x}_1\hat{w}_2 - \hat{x}_2\hat{w}_3] \\ -b\,[\wp(\hat{x}_1, \hat{x}_2)\hat{w}_1 + \hat{x}_2\hat{w}_2 - \hat{x}_1\hat{w}_3)] \end{pmatrix}.$$

It is globally defined and globally convergent for any solution of the oscillator initialized in the set $\mathcal{X}$ given in (8.3). Results of a simulation are given in Fig. 11.1. Notice that the observer converges despite the fact that $\hat{x}_1$ and $\hat{x}_2$ are initialized at the singularity. This would not have been possible with observer (9.6), i.e., without adding the fictitious output. By the way, observe that w_2 and w_3 present a violent peak at $t = 0$. This is due to the fact that $\hat{x}_1$ and $\hat{x}_2$ are around the singularity, where only the fictitious output (which has a very small but nonzero value) preserves the invertibility of the Jacobian. We used a step-variable integration scheme to take this into account. ▲

Remark 11.1 It is interesting to notice that the manifold $\hat{\xi}_5 = \hat{\xi}_6 = 0$ is invariant. If $\hat{w}$ is initialized at 0 (which is reasonable since its *true* value is 0), $(\hat{\xi}_5(0), \hat{\xi}_6(0)) = T_{56}(x_0) = 0$, and therefore, $\hat{\xi}_5$ and $\hat{\xi}_6$ remain 0 along the observer trajectories. This questions the interest of having added those two dimensions and actually implies the existence of an observer with dimension reduced to 4. However, it is shown in [2] that such an observer cannot be expressed with coordinates $(x, \bar{w})$ in $\mathbb{R}^4$.

11.1.3 A Universal Complementation Method

In the previous example, the Jacobian complementation is made possible by increasing d_ξ, i.e., augmenting the number of coordinates of T. Actually, if we augment T with d_x zeros, the possibility of a Jacobian complementation is guaranteed. Indeed, the full-rank matrix $\begin{pmatrix} \frac{\partial T}{\partial x}(x) \\ 0 \end{pmatrix}$ can always be complemented[2] with

$$\gamma = \begin{pmatrix} -I \\ \frac{\partial T}{\partial x}(x)^\top \end{pmatrix} .$$

This follows from the identity (Schur complement) involving invertible matrices

$$\begin{pmatrix} \frac{\partial T}{\partial x}(x) & -I \\ 0 & \frac{\partial T}{\partial x}(x)^\top \end{pmatrix} \begin{pmatrix} 0 & I \\ I & \frac{\partial T}{\partial x}(x) \end{pmatrix} = \begin{pmatrix} -I & 0 \\ \frac{\partial T}{\partial x}^\top(x) & \frac{\partial T}{\partial x}(x)^\top \frac{\partial T}{\partial x}(x) \end{pmatrix} .$$

So we have here a universal method to solve Problem 8.1. Its drawback is that the dimension of the state increases by d_ξ, instead of $d_\xi - d_x$.

11.2 A Global Example: Luenberger Design for the Oscillator

Let us now come back to the Luenberger observer presented in Sect. 8.1.1.2 for the oscillator with unknown frequency. Although an inversion of the transformation was proposed in [6] based on the resolution of a minimization problem, we want to show here how this step can be avoided.

Recall that the transformation is given by

$$T(x) = \left(\frac{\lambda_1 x_1 - x_2}{\lambda_1^2 + x_3}, \frac{\lambda_2 x_1 - x_2}{\lambda_2^2 + x_3}, \frac{\lambda_3 x_1 - x_2}{\lambda_3^2 + x_3}, \frac{\lambda_4 x_1 - x_2}{\lambda_4^2 + x_3} \right)$$

and its Jacobian

$$\frac{\partial T}{\partial x}(x) = \begin{pmatrix} \frac{\lambda_1}{\lambda_1^2+x_3} & -\frac{1}{\lambda_1^2+x_3} & -\frac{T_1(x)}{\lambda_1^2+x_3} \\ \frac{\lambda_2}{\lambda_2^2+x_3} & -\frac{1}{\lambda_2^2+x_3} & -\frac{T_2(x)}{\lambda_2^2+x_3} \\ \frac{\lambda_3}{\lambda_3^2+x_3} & -\frac{1}{\lambda_3^2+x_3} & -\frac{T_3(x)}{\lambda_3^2+x_3} \\ \frac{\lambda_4}{\lambda_4^2+x_3} & -\frac{1}{\lambda_4^2+x_3} & -\frac{T_4(x)}{\lambda_4^2+x_3} \end{pmatrix} .$$

[2] Actually, I can be replaced by any C^1 function B the values of which are $d_\xi \times d_\xi$ matrices with positive definite symmetric part.

The complementation is quite easy because there is only one dimension to add: We could just add a column $\gamma(x)$ consisting of the corresponding minors as suggested in Sect. 9.2. However, this would produce a diffeomorphism on $\mathscr{X} \times B_\varepsilon(0)$ for some ε, where $\mathscr{X}$ defined in (8.3) is not contractible due to the observability singularity at $x_1 = x_2 = 0$. Therefore, no image extension is possible and it would be necessary to ensure that $\hat{w}$ remains small and $(\hat{x}_1, \hat{x}_2)$ far from $(0, 0)$ by some other means. As in Example 11.2, we first try to remove this singularity.

Again, we assume the system solutions remain in $\mathscr{X}$ and add the same fictitious output y_2 as before, which vanishes in $\mathscr{X}$ and which is nonzero when (x_1, x_2) is close to the origin, namely

$$y_2 = \wp(x_1, x_2)x_3 ,$$

where $\wp$ is defined in (11.1). Once again, it is possible to show[3] that by adding y_2 to T, the observability singularity disappears; namely, the function

$$T(x) = \left(\frac{\lambda_1 x_1 - x_2}{\lambda_1^2 + x_3}, \frac{\lambda_2 x_1 - x_2}{\lambda_2^2 + x_3}, \frac{\lambda_3 x_1 - x_2}{\lambda_3^2 + x_3}, \frac{\lambda_4 x_1 - x_2}{\lambda_4^2 + x_3}, \wp(x_1, x_2)x_3\right)$$

is an injective immersion on

$$\tilde{\mathscr{S}} = \mathbb{R}^2 \times \mathbb{R}_+ .$$

Although the Jacobian complementation problem is solvable for this T according to Wazewski's Theorem 9.4 because $\tilde{\mathscr{S}}$ is contractible, we want to avoid the lengthy computations entailed by this method. We are going to see in the following that it is possible if one rather takes

$$T(x) = \left(\frac{\lambda_1 x_1 - x_2}{\lambda_1^2 + x_3}, \frac{\lambda_2 x_1 - x_2}{\lambda_2^2 + x_3}, \frac{\lambda_3 x_1 - x_2}{\lambda_3^2 + x_3}, \frac{\lambda_4 x_1 - x_2}{\lambda_4^2 + x_3}, \wp(x_1, x_2)x_3, 0\right) \quad (11.6)$$

which is still an injective immersion on $\tilde{\mathscr{S}}$. The new Jacobian takes the form

[3] In [6], it is shown that for any $r > 0$, there exists $L_r > 0$ such that for all (x_a, x_b) in $\mathbb{R}^2 \times (0, r)$, $|x_{1,a} - x_{1,b}| + |x_{2,a} - x_{2,b}| + \frac{x_{1,a}+x_{1,b}+x_{2,a}+x_{2,b}}{2}|x_{3,a} - x_{3,b}| \leq L_r|T_{14}(x_a) - T_{14}(x_b)|$ where T_{14} denotes the first four components of T. Therefore, $T_{14}(x_a) = T_{14}(x_b)$ implies that $x_{1,a} = x_{1,b}$ and $x_{2,a} = x_{2,b}$: Either one of them is nonzero, and in that case, the inequality says that we have also $x_{3,a} = x_{3,b}$, or they are all zero but then $T_5(x_a) = T_5(x_b)$ implies that $x_{3,a} = x_{3,b}$. We conclude that T is injective on $\tilde{\mathscr{S}}$. Now, applying the inequality between x and $x + hv$ and making h go to zero, we get that $\frac{\partial T_{14}}{\partial x}(x)v = 0$ implies that $v_1 = v_2 = 0$ and $v_3 = 0$ if either x_1 or x_2 is nonzero. If they are both zero, $\frac{\partial T_5}{\partial x}(x)v = 0$ with $v_1 = v_2 = 0$ gives $v_3 = 0$. Thus, $\frac{\partial T}{\partial x}(x)$ is full-rank.

$$
\frac{\partial T}{\partial x}(x) = \begin{pmatrix} \frac{\lambda_1}{\lambda_1^2+x_3} & -\frac{1}{\lambda_1^2+x_3} & -\frac{T_1(x)}{\lambda_1^2+x_3} \\ \frac{\lambda_2}{\lambda_2^2+x_3} & -\frac{1}{\lambda_2^2+x_3} & -\frac{T_2(x)}{\lambda_2^2+x_3} \\ \frac{\lambda_3}{\lambda_3^2+x_3} & -\frac{1}{\lambda_3^2+x_3} & -\frac{T_3(x)}{\lambda_3^2+x_3} \\ \frac{\lambda_4}{\lambda_4^2+x_3} & -\frac{1}{\lambda_4^2+x_3} & -\frac{T_4(x)}{\lambda_4^2+x_3} \\ A(x) & B(x) & C(x) \\ 0 & 0 & 0 \end{pmatrix},
$$

for some appropriate functions A, B, and C. We now have to add two columns. Let us first simplify the matrix to be complemented by noticing that

$$
M(x_3, \lambda_i)\,\frac{\partial T}{\partial x}(x) = \begin{pmatrix} 1 & 0 & m_1(x) \\ 0 & 1 & m_2(x) \\ 0 & 0 & m_3(x) \\ 0 & 0 & m_4(x) \\ A(x) & B(x) & C(x) \\ 0 & 0 & 0 \end{pmatrix} \tag{11.7}
$$

where $M(x_3, \lambda_i)$ is the invertible matrix

$$
M(x_3, \lambda_i) = \begin{pmatrix} \mathfrak{D}^{-1}(\lambda_i) & 0_{4\times 2} \\ 0_{2\times 4} & I_{2\times 2} \end{pmatrix} \begin{pmatrix} \lambda_1^2 + x_3 & 0 & 0 & 0 & 0 & 0 \\ 0 & \lambda_2^2 + x_3 & 0 & 0 & 0 & 0 \\ 0 & 0 & \lambda_3^2 + x_3 & 0 & 0 & 0 \\ 0 & 0 & 0 & \lambda_4^2 + x_3 & 0 & 0 \\ 0 & 0 & 0 & 0 & 1 & 0 \\ 0 & 0 & 0 & 0 & 0 & 1 \end{pmatrix}
$$

and $\mathfrak{D}(\lambda_i)$ is an appropriate Vandermonde matrix associated to the λ_i. So now we are left with complementing the matrix given by (11.7). Observing that right-multiplying (11.7) by the invertible matrix $N(x) = \begin{pmatrix} 1 & 0 & -m_1(x) \\ 0 & 1 & -m_2(x) \\ 0 & 0 & 1 \end{pmatrix}$ gives

$$
\begin{pmatrix} 1 & 0 & 0 \\ 0 & 1 & 0 \\ 0 & 0 & m_3(x) \\ 0 & 0 & m_4(x) \\ A(x) & B(x) & m_5(x) \\ 0 & 0 & 0 \end{pmatrix},
$$

with

$$
m_5(x) = C(x) - m_1(x)A(x) - m_2(x)B(x)\,,
$$

we conclude first that the vector $(m_3(x), m_4(x), m_5(x), 0)$ in $\mathbb{R}^4$ is nonzero on $\tilde{\mathscr{S}}$ and then that (11.7) can be simply complemented by complementing the vector

$(m_3(x), m_4(x), m_5(x), 0)$ into an invertible 4×4 matrix. Note that this is the solvable problem $\tilde{P}[4, 1]$ from (9.7), while without adding the zero output y_3, we would have obtained $\tilde{P}[3, 1]$ which is not solvable. An explicit solution to $\tilde{P}[4, 1]$ is given in Sect. 9.2, but we can here also exploit the very particular structure of the vector and use the remark made in Sect. 11.1.3 that the matrix

$$P(x) = \begin{pmatrix} m_3(x) & -1 & 0 & 0 \\ m_4(x) & 0 & -1 & 0 \\ m_5(x) & 0 & 0 & -1 \\ 0 & m_3(x) & m_4(x) & m_5(x) \end{pmatrix}$$

is invertible as soon as $(m_3(x), m_4(x), m_5(x))$ is nonzero.

Reversing the transformations, we thus manage to extend the Jacobian of T into a matrix of dimension 6 whose determinant is nonzero on $\tilde{\mathcal{S}}$. Adding three state components to the system state, we obtain a diffeomorphism T_a on $\mathcal{S}_a = \tilde{\mathcal{S}} \times B_\varepsilon$, with ε sufficiently small. As in Example 11.2, we add two exponentially decaying dynamics for the added dimensions $\hat{\xi}_5$ and $\hat{\xi}_6$, thus leading to the observer

$$\begin{pmatrix} \dot{\hat{x}} \\ \dot{\hat{w}} \end{pmatrix} = \left(\frac{\partial T_a}{\partial x}(\hat{x}, \hat{w}) \right)^{-1} \left(A T_a(\hat{x}, \hat{w}) + B y_1 \right) \tag{11.8}$$

where $B = [1, 1, 1, 1, 0, 0]^\top$, $A = \mathtt{diag}(-\lambda_1, -\lambda_2, -\lambda_3, -\lambda_4, -a, -b)$, and a and b are two positive real numbers. The expression of the Jacobian of the extended function is omitted here due to its complexity, but it can be obtained by straightforward symbolic computations.

The singularity at $(\hat{x}_1, \hat{x}_2) = 0$ has disappeared, but we still need to ensure that $\hat{x}_3$ remains positive, or at least greater than $-\min\{\lambda_i^2\}$. Besides, unlike the high-gain observer (11.5), the invertibility of the extended Jacobian is only guaranteed for w in B_ε. To make sure the solutions remain in $\mathcal{S}_a = \tilde{\mathcal{S}} \times B_\varepsilon$, we should solve Problem 8.2, namely extending T_a into a diffeomorphism T_e whose image of $\mathcal{S}_a$ covers $\mathbb{R}^6$. Since $\mathcal{S}_a$ is diffeomorphic to $\mathbb{R}^6$, we know it is theoretically possible by Theorem 10.1 and replacing T_a by the new surjective diffeomorphism T_e in (11.8) would give an observer whose solutions are ensured to exist for all t.

Unfortunately, due to the complexity of the expression of T_a, we are not yet able to achieve such an extension. The consequence is that there may exist a set of initial conditions and parameters such that the corresponding trajectory of observer (11.8) encounters a singularity of the jacobian of T_a and thus diverges. A way of reducing this set is to approximate the image of $T_a(\mathcal{S}_a)$, as proposed in Example 10.1. In the present case, we have (denoting T_{14} the first four components of T defined in (11.6)),

$$(\xi_1, \xi_2, \xi_3, \xi_4) = T_{14}(x) \quad \Longleftrightarrow \quad \begin{pmatrix} \lambda_1^2 \xi_1 & -\lambda_1 & 1 & \xi_1 \\ \lambda_2^2 \xi_2 & -\lambda_2 & 1 & \xi_2 \\ \lambda_3^2 \xi_3 & -\lambda_3 & 1 & \xi_3 \\ \lambda_4^2 \xi_4 & -\lambda_4 & 1 & \xi_4 \end{pmatrix} \begin{pmatrix} 1 \\ x_1 \\ x_2 \\ x_3 \end{pmatrix} = 0$$

and thus

$$F(T(x)) = F(T_a(x, 0)) = 0$$

where F is the quadratic function defined by

$$F(\xi) = \det \begin{pmatrix} \lambda_1^2 \xi_1 & -\lambda_1 & 1 & \xi_1 \\ \lambda_2^2 \xi_2 & -\lambda_2 & 1 & \xi_2 \\ \lambda_3^2 \xi_3 & -\lambda_3 & 1 & \xi_3 \\ \lambda_4^2 \xi_4 & -\lambda_4 & 1 & \xi_4 \end{pmatrix} + \xi_5^2 + \xi_6^2 \,.$$

Therefore, replacing $\xi^\top M \xi$ by $F(\xi)$ in (10.2)–(10.7) gives a diffeomorphism ϕ from $\mathbb{R}^6$ to

$$E = \left\{ \xi \in \mathbb{R}^6 : \ F(\xi)^2 < \delta \right\}$$

and taking $T_e = \phi^{-1} \circ T_a$ instead of T_a ensures that for any observer solution $t \mapsto \hat{\xi}(t)$, our estimate defined by $(\hat{x}, \hat{w}) = T_e^{-1}(\hat{\xi})$ will be such that $T_a(\hat{x}, \hat{w})$ remains in

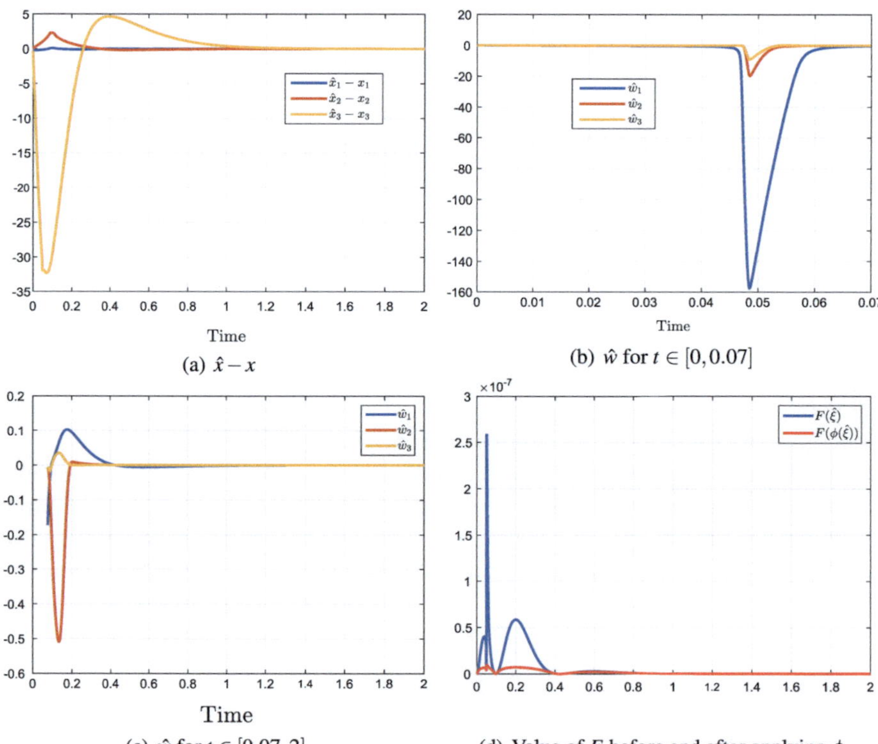

(a) $\hat{x} - x$

(b) $\hat{w}$ for $t \in [0, 0.07]$

(c) $\hat{w}$ for $t \in [0.07, 2]$

(d) Value of F before and after applying ϕ

Fig. 11.2 Luenberger observer (11.8) with $\hat{x}_1 = 0.08, \hat{x}_2 = \hat{x}_3 = 0, \lambda_1 = 6, \lambda_2 = 9, \lambda_3 = 14, \lambda_4 = 15, a = b = 1$, and $T_e = \phi^{-1} \circ T_a$ instead of T_a for the oscillator (8.2) initialized at (0.35, 0, 0.4). The simulation was done with a variable step Euler algorithm

E. When δ goes to zero, E gets closer to $T_a(\tilde{\mathscr{S}} \times \{0\})$ and thus we can hope that $\hat{w}$ will remain sufficiently small to keep the invertibility of the Jacobian of T_e. We indeed observe in simulations that taking T_e instead of T_a enables to ensure completeness of some of the solutions which otherwise diverge with T_a. An example is given in Fig. 11.2: Before $t = 0.05$, the observer trajectory is close to a singularity, $\hat{w}$ tends to become very large (see Fig. 11.2b), so does $F(\hat{\xi})$, but ϕ enables to reduce $F(\hat{\xi})$ (see Fig. 11.2d) and thus prevent $\hat{w}$ from becoming too large and encounter the singularity. Unfortunately, although the set of initial conditions leading to uncomplete solutions is significantly reduced by this method, it does not completely disappear.

11.3 Generalization to a Time-Varying T

In Assumption 8.1, it is supposed that the transformation T from the given x-coordinates to the ξ-coordinates is stationary. But we have seen in Part II that it is sometimes easier/necessary to consider a time-varying transformation which depends on the input, and apply Theorem 1.1. It is thus legitimate to wonder if the methodology presented in this part is still useful. In fact, the same tools can be applied in the sense that:

- Assumption 8.1 should now provide for each u in $\mathscr{U}$ a C^1 function $T : \mathbb{R}^{d_x} \times \mathbb{R} \to \mathbb{R}^{d_\xi}$, subsets $\mathscr{S}_t$ and $\mathscr{X}_t$ of $\mathbb{R}^{d_x}$ and a set $\varphi\mathscr{T}$ of pairs $(\varphi, \mathscr{T})$ such that:
  - For all t in $[0, +\infty)$, $x \mapsto T(x, t)$ is an injective immersion on $\mathscr{S}_t$.
  - For all x_0 in $\mathscr{X}_0$ and all t in $[0, +\infty)$, $X(x_0; t; u)$ is in $\mathscr{X}_t$.
  - For all x in $\mathscr{X}_t$, $\mathscr{T}(T(x, t), t) = x$ and φ is such that the appropriate convergence in the ξ-coordinates is achieved.
- Problems 8.1 and 8.2 can then be solved applying the tools of Chaps. 9 and 10 on $x \mapsto T(x, t)$ for each t. This leads to a function $T_e : \mathbb{R}^{d_x} \times \mathbb{R}^{d_\xi - d_x} \times \mathbb{R} \to \mathbb{R}^{d_\xi}$ and open subsets $\mathscr{S}_{a,t}$ of $\mathbb{R}^{d_\xi}$ containing $\mathscr{X} \times \{0\}$ such that for all t in $[0, +\infty)$, $(x, w) \mapsto T_e(x, w, t)$ is a diffeomorphism on $\mathscr{S}_{a,t}$ verifying:

$$T_e(x, 0, t) = T(x, t) \qquad \forall x \in \mathscr{X}_t \tag{11.9}$$

  and

$$T_e(\mathscr{S}_{a,t}, t) = \mathbb{R}^{d_\xi} . \tag{11.10}$$

- In order to ensure

$$\dot{\overbrace{T_e(\hat{x}, \hat{w}, t)}} = \varphi(T_e(\hat{x}, \hat{w}, t), \hat{x}, u, y) ,$$

  and conclude as before that

$$\lim_{t \to +\infty} \left| T_e\left(\hat{X}(\hat{x}_0, \hat{w}_0; t; u), \hat{W}(\hat{x}_0, \hat{w}_0; t; u), t\right) - T(X(x_0; t; u), t) \right| = 0 , \tag{11.11}$$

we must take into account the dependence of T_e on t and take:

$$\dot{\overbrace{\begin{bmatrix} \hat{x} \\ \hat{w} \end{bmatrix}}} = \left(\frac{\partial T_e}{\partial(\hat{x},\hat{w})}(\hat{x},\hat{w},t)\right)^{-1}\left(\varphi(T_e(\hat{x},\hat{w},t),\hat{x},u,y) - \frac{\partial T_e}{\partial t}(\hat{x},\hat{w},t)\right) . \tag{11.12}$$

- Finally, to conclude from (11.11) that $\hat{x}$ converges to x and $\hat{w}$ to 0, we further need that the injectivity of $(x, w) \mapsto T_e(x, w, t)$ be uniform in t. When the dependence on t of T_e comes from the input (and its derivatives), this property is often satisfied, in particular when those signals are bounded in time (see Lemma A.12). Note that a special attention should also be given to the set $\mathscr{S}_{a,t}$ which could be of the form $\mathscr{S}_t \times B_{\varepsilon(t)}$ with ε going to 0 with t. Thus, it should be checked that ε is lower bounded. A justification as to why this should be true in practice appears in the next section.

We give in the following section some elements of justification, and then, we illustrate this on an example about aircraft landing.

11.3.1 Partial Theoretical Justification

Suppose that for all t in $[0, +\infty)$, $x \mapsto T(x, t)$ is an injective immersion on some open set $\mathscr{S}_t$. Consider the extended system

$$\begin{cases} \dot{x} = f(x, u(t)) \\ \dot{t} = 1 \end{cases} \quad , \quad \underline{y} = \begin{pmatrix} h(x, u(t)) \\ t \end{pmatrix}$$

with state $\underline{x} = (x, t)$. Then, the function

$$\underline{T}(\underline{x}) = (T(x, t), t)$$

is an injective immersion on

$$\underline{\mathscr{S}} = \{(x, t) \in \mathbb{R}^{d_x} \times [0, +\infty) \ : \ x \in \mathscr{S}_t\}$$

and complementing its Jacobian

$$\frac{\partial \underline{T}}{\partial \underline{x}}(\underline{x}) = \begin{pmatrix} \frac{\partial T}{\partial x}(x, t) & \frac{\partial T}{\partial t}(x, t) \\ 0 & 1 \end{pmatrix}$$

on $\underline{\mathscr{S}}$ is equivalent to complementing that of $x \mapsto T(x, t)$ on $\mathscr{S}_t$ for each t. Indeed, if $\gamma(x, t)$ is a complementation of $\frac{\partial T}{\partial x}(x, t)$ on $\mathscr{S}_t$ for each t, $\underline{\gamma}(\underline{x}) = \begin{pmatrix} \gamma(x, t) \\ 0 \end{pmatrix}$ is a complementation for $\frac{\partial \underline{T}}{\partial \underline{x}}(\underline{x})$ on $\underline{\mathscr{S}}$. And conversely, if $\underline{\gamma}(\underline{x}) = \begin{pmatrix} \gamma(x, t) \\ \alpha \end{pmatrix}$ complements $\frac{\partial \underline{T}}{\partial \underline{x}}(\underline{x})$, then $\gamma(x, t) - \frac{\partial T}{\partial t}(x, t)$ complements $\frac{\partial T}{\partial x}(x, t)$.

We conclude that it is not restrictive to look for a complementation of the Jacobian of $x \mapsto T(x, t)$ at each time t. Assume it has been done and take

$$\underline{\gamma}(\underline{x}) = \begin{pmatrix} \gamma(x,t) \\ 0 \end{pmatrix} .$$

Following the methodology, we consider

$$\underline{T_a}(\underline{x}, w) = \underline{T}(\underline{x}) + \underline{\gamma}(\underline{x})w = \begin{pmatrix} T(x,t) + \gamma(x,t)w \\ t \end{pmatrix} = \begin{pmatrix} T_a(x,w,t) \\ t \end{pmatrix} .$$

Beware that Lemma 9.1 does not apply directly because $\underline{\mathscr{S}}$ is not bounded; thus, we cannot directly conclude that there exists $\varepsilon > 0$ such that $\underline{T_a}$ is a diffeomorphism on $\underline{\mathscr{S}_a} = \underline{\mathscr{S}} \times B_\varepsilon$. However, the reader may check in the proof of [1, Proposition 2] that if $\frac{\partial T}{\partial \underline{x}}(x,t)$, $\gamma(x,t)$ and $\frac{\partial \gamma}{\partial \underline{x}}(x,t)$ are bounded on $\underline{\mathscr{S}}$, the Jacobian of $\underline{T_a}$ is full-rank on $\underline{\mathscr{S}_a}$ for some ε sufficiently small. This condition is often verified in practice when the inputs are bounded. It follows that we can reasonably assume Problem 8.1 solved, and leaving aside Problem 8.2, this leads to an observer of the type (denoting $\underline{T_a}(x,w,t)$ rather than $\underline{T_a}(x,t,w)$)

$$\dot{\overbrace{\begin{bmatrix} \hat{x} \\ \hat{w} \\ \hat{t} \end{bmatrix}}} = \left(\frac{\partial \underline{T_a}}{\partial (x,w,t)}(\hat{x},\hat{w},\hat{t}) \right)^{-1} \begin{pmatrix} \varphi(T_a(\hat{x},\hat{w},\hat{t}),\hat{x},u,\underline{y}) \\ \varphi_1(\hat{t},\underline{y}) \end{pmatrix}$$

where φ_1 should be an observer for t and we have

$$\left(\frac{\partial \underline{T_a}}{\partial (x,w,t)}\right)^{-1} = \begin{pmatrix} \frac{\partial T_a}{\partial (x,w)} & \frac{\partial T_a}{\partial t} \\ 0 & 1 \end{pmatrix}^{-1} = \begin{pmatrix} \left(\frac{\partial T_a}{\partial (x,w)}\right)^{-1} & -\left(\frac{\partial T_a}{\partial (x,w)}\right)^{-1} \frac{\partial T_a}{\partial t} \\ 0 & 1 \end{pmatrix} .$$

Of course, t being well known without any noise, we can replace $\hat{t}$ by t and φ_1 by the constant function 1. This finally gives the "reduced-order" observer (11.12).

11.3.2 Application to Image-Based Aircraft Landing

In [4, 5], the authors use image processing to estimate the deviations of an aircraft with respect to the runaway during a landing operation thanks to vision sensors such as cameras and inertial sensors embarked on the aircraft. The objective is to make landing possible without relying on external technologies or any knowledge about the runaway. In order to estimate the position of the plane, the idea is to follow the change of position of particular points and/or particular lines on the images provided by the cameras. A strategic choice of those points/lines must be made in order to guarantee observability during the whole duration of the landing operation: For instance, a point may disappear from the image, and a line can stop moving on the image in some particular alignment conditions, thus providing no (or only partial) information about the movement of the aircraft. A full study of those methods can be found in [3]. A possible choice ensuring observability is to follow on the image the position of the two lateral lines of the runaway and the reference point at the end of the runaway. It gives the following model:

$$\left\{\begin{aligned} \dot\theta_1 &= \sigma_1(\rho_1,\theta_1,t) + \pi_1(\rho_1,\theta_1,t)\eta \\ \dot\rho_1 &= \sigma_2(\rho_1,\theta_1,t) + \pi_2(\rho_1,\theta_1,t)\eta \\ \dot\theta_2 &= \sigma_1(\rho_2,\theta_2,t) + \pi_1(\rho_2,\theta_2,t)\eta \\ \dot\rho_2 &= \sigma_2(\rho_2,\theta_2,t) + \pi_2(\rho_2,\theta_2,t)\eta \\ \dot\nu_1 &= (V_H\nu_1 - V_X)\eta \\ \dot\nu_2 &= (V_H\nu_2 - V_Y)\eta \\ \dot\eta &= V_H\eta^2 \end{aligned}\right. \quad , \quad y = (\theta_1,\rho_1,\theta_2,\rho_2,\nu_1,\nu_2)$$

where (θ_i,ρ_i) and (ν_1,ν_2) are the measured position on the image of the two lines and the point, respectively; the functions σ and π are defined by

$$\begin{aligned} \sigma_1(\rho,\theta,t) &= -\omega_1\rho\cos\theta - \omega_2\rho\sin\theta - \omega_3 \\ \sigma_2(\rho,\theta,t) &= (1+\rho^2)(\omega_1\sin\theta - \omega_2\cos\theta) \\ \pi_1(\rho,\theta,t) &= (a\sin\theta - b\cos\theta)(v_1\cos\theta + v_2\sin\theta - v_3\rho) \\ \pi_2(\rho,\theta,t) &= (a\rho\cos\theta + b\rho\sin\theta + c)(v_1\cos\theta + v_2\sin\theta - v_3\rho) \end{aligned}$$

where the aircraft velocities v and ω expressed in the camera frame, the aircraft velocities V_X, V_Y, and V_H expressed in the runway frame, and camera orientations (a,b,c) are known input signals.

Denoting $x_m = (\theta_1,\rho_1,\theta_2,\rho_2,\nu_1,\nu_2)$ the measured part of the state, we obtain a model with state

$$x = (x_m,\eta) \in \mathbb{R}^7$$

and dynamics of the form[4]

$$\left\{\begin{aligned} \dot x_m &= \Sigma(x_m,t) + \Pi(x_m,t)\,\eta \\ \dot\eta &= V_H(t)\,\eta^2 \end{aligned}\right. \quad , \quad y = x_m \ , \tag{11.13}$$

where the action of the input $u = (a,b,c,v,w,V_X,V_Y,V_H)$ is represented by a time dependence[5] to simplify the notations in the rest of this section. This system is observable if and only if the unmeasured state η can be uniquely determined from the

---

[4]Note that whatever the number of chosen lines and points in the image, the model can always be written in this form, only the dimensions of x_m and the input change.

[5]This comes back to choosing one particular input law, but the reader may check that the same design works for any input such that the observability assumption and the saturation by $\overline{\Phi}$ in (11.18) are valid.

knowledge of the measured state x_m. From the structure of the dynamics, we notice that this is possible if the quantity

$$\delta(x_m, t) = \Pi(x_m, t)^\top \Pi(x_m, t) \tag{11.14}$$

never vanishes. It is the case in practice, thanks to a sensible choice of lines and point (see [3] for a thorough observability analysis during several landing operations).

11.3.2.1 A High-Gain Observer

Assumption 11.1

- The input signal $u = (v, w, V_X, V_Y, V_H)$ and its first derivative are bounded in time.
- There exists a strictly positive number ε and a compact subset $\mathscr{C}$ of $\mathbb{R}^7$ such that for any x_0 in $\mathscr{X}_0$, the corresponding solution to System (11.13) verifies

$$X(x_0; t; u) \in \mathscr{X}_t = \mathscr{S}_t \cap \mathscr{C} \quad \forall t \in [0, +\infty) \tag{11.15}$$

where for each time t, we define

$$\mathscr{S}_t = \{x \in \mathbb{R}^7 \,,\; \delta(x_m, t) \geq \varepsilon\} \,.$$

In other words, δ remains greater than ε along any solution of the system, making it observable. Under this assumption, we know that the state x can be reconstructed from the measurement x_m and its first derivative. We thus consider the transformation $T_0 : \mathbb{R}^7 \times \mathbb{R} \to \mathbb{R}^{12}$ made of y and its first derivative, i.e.,

$$T_0(x, t) = \overline{\mathbf{H}}_2(x, u(t)) = \begin{pmatrix} x_m \\ \Sigma(x_m, t) + \Pi(x_m, t)\, \eta \end{pmatrix} . \tag{11.16}$$

For any t in $\mathbb{R}$, $T_0(\cdot\,,\ t)$ is an injective immersion on $\mathscr{S}_t$. Since u, $\dot{u}$ and the trajectories are bounded, we deduce from Theorem 7.1 and Remark 7.1 that T_0 transforms the system into a phase-variable form

$$\begin{cases} \dot{\xi}_m = \xi_d \\ \dot{\xi}_d = \Phi_2(\xi, u(t), \dot{u}(t)) \end{cases} , \quad y = \xi_m \tag{11.17}$$

where ξ_m denotes the first six components of ξ and ξ_d the six others, and Φ_2 can be defined by

$$\Phi_2(\xi, \nu_0, \nu_1) = \mathsf{sat}_{\overline{\Phi}}(L_f^2 \hbar(\mathscr{T}_0(\xi, t), \nu_0, \nu_1)) \tag{11.18}$$

with $\overline{f}$ and $\hbar$ as defined in Definition 5.1, $\overline{\Phi}$ a bound of $L^2_{\overline{f}}\hbar(x, \nu_0, \nu_1)$ for x in $\mathscr{C}$ and (ν_0, ν_1) bounded by the bound for $(u, \dot{u})$, and $\xi \mapsto \mathscr{T}_0(\xi, \cdot)$ any locally Lipschitz function defined on $\mathbb{R}^{12}$ such that it is a left inverse[6] of $x \mapsto T_0(x, \cdot)$ for x in $\mathscr{X}_t$.

We have the following observer for System (11.17):

$$\begin{cases} \dot{\hat{\xi}}_m = \hat{\xi}_d + Lk_1(y - \hat{\xi}_m) \\ \dot{\hat{\xi}}_d = \Phi_2(\hat{\xi}, u, \dot{u}) + L^2 k_2(y - \hat{\xi}_m) \end{cases} , \quad y = \xi_m \tag{11.19}$$

with $k_1, k_2 > 0$ and L sufficiently large. Although a left inverse $\mathscr{T}_0$ of T_0 can be found in that case, and an estimate $\hat{x}$ of x could be computed by $\hat{x} = \mathscr{T}_0(\hat{\xi}, t)$ as proposed by Theorem 1.1, we would like to express the dynamics of this observer directly in the x-coordinates.

11.3.2.2 Observer in the Given Coordinates

Fictitious output Following the same idea as for the oscillator with unknown frequency, we start by removing the injectivity singularity of T_0 outside of $\mathscr{S}_t$; i.e., we look for an alternative function T which is an injective immersion on $\mathbb{R}^7$. Notice that the function

$$\wp(x_m, t) = \max\left\{\varepsilon - \delta(x_m, t), 0\right\}^4 \tag{11.20}$$

is zero in $\mathscr{S}_t$ and nonzero outside of $\mathscr{S}_t$. According to (11.15), this function remains equal to 0 along the solutions and therefore so does the fictitious output

$$y_7 = \wp(x_m, t)\eta \; .$$

It follows that y_7 can be considered as an extra measurement traducing the information of observability. Consider now the function

$$T(x, t) = (T_0(x, t)\, ,\, \wp(x_m, t)\eta) \; .$$

Unlike $T_0(\cdot, t)$, $T(\cdot, t)$ is an injective immersion on the whole space $\mathbb{R}^7$ for all t. Indeed, $\Pi(x_m, t)$ and $\wp(x_m, t)$ cannot be zero at the same time so that the new coordinate $\wp(x, t)\eta$ enables to have the information on η when Π is zero. Besides, its Jacobian

$$\frac{\partial T}{\partial x}(x, t) = \begin{pmatrix} I_{6\times 6} & 0_{6\times 1} \\ * & \Pi(x_m, t) \\ * & \wp(x_m, t) \end{pmatrix} \tag{11.21}$$

is full-rank everywhere.

[6]Take for instance $\mathscr{T}_0(\xi, t) = \left(\xi_m, \frac{\Pi(\xi_m)^T(\xi_d - \Sigma(\xi_m, t))}{\max\{\delta(\xi_m, t), \varepsilon\}}\right)$.

Immersion augmentation into diffeomorphism by Jacobian complementation. Following the methodology presented in this Part III, we extend the injective immersion $T(\cdot, t)$ into a diffeomorphism. The first step consists in finding a C^1 matrix $\gamma(x, t)$ in $\mathbb{R}^{13\times 6}$ such that the matrix

$$\begin{pmatrix} \frac{\partial T}{\partial x}(x, t) & , & \gamma(x, t) \end{pmatrix}$$

is invertible for any x and any t. In others words, we want to complement the full-rank rectangular matrix $\frac{\partial T}{\partial x}(x, t)$ with six vectors in $\mathbb{R}^{13}$ which make it square and invertible. Thanks to the identity block, it is in fact sufficient to find six independent vectors in $\mathbb{R}^7$ which complement the vector $\begin{pmatrix} \Pi(x_m, t) \\ \wp(x_m, t) \end{pmatrix}$. A first solution would be to implement Wazewski's algorithm on $\mathbb{R}^6$ which is contractible, like in Example 9.3, but this leads to rather tedious computations. Since Problem $\tilde{P}[7, 1]$ is not in the list (9.7) of cases admitting universal formulas, we could had another fictitious output $y_8 = 0$ like we did for the oscillator to recover a solvable problem $\tilde{P}[8, 1]$. We present here another path which does not necessitate lengthy computations nor an additional output. The idea comes from the remark made in Sect. 11.1.3 that when $\begin{pmatrix} \frac{\partial T}{\partial x} \\ 0 \end{pmatrix}$ is full-rank, it can always be complemented by $\begin{pmatrix} -I \\ \frac{\partial T}{\partial x}^\top \end{pmatrix}$ because the resulting matrix has a determinant equal to $\det\left(\frac{\partial T}{\partial x}^\top \frac{\partial T}{\partial x}\right) \neq 0$. In our case, we remark that the determinant of the matrix $\begin{pmatrix} \Pi(x_m, t) & -I_{6\times 6} \\ \wp(x_m, t) & \Pi(x_m, t)^\top \end{pmatrix}$ is equal to $\wp(x_m, t) + \Pi(x_m, t)^T \Pi(x_m, t)$ which never vanishes by definition. Thus, a possible candidate for complementation is:

$$\gamma(x_m, t) = \begin{pmatrix} 0_{6\times 6} \\ -I_{6\times 6} \\ \Pi(x_m, t)^\top \end{pmatrix} .$$

As recommended by Lemma 9.1, we now introduce the extension of T defined on $\mathbb{R}^7 \times \mathbb{R}^6 \times \mathbb{R}$ by

$$T_e(x, w, t) = T(x, t) + \gamma(x_m, t)w . \tag{11.22}$$

Besides, thanks to the fact that γ does not depend on η, we have:

$$\frac{\partial T_e}{\partial(x, w)}(x, w, t) = \begin{pmatrix} \mathsf{Id}_{6\times 6} & 0_{6\times 1} & 0_{6\times 6} \\ * & \Pi(x_m, t) & -\mathsf{Id}_{6\times 6} \\ * & \wp(x_m, t) & \Pi(x_m, t)^\top \end{pmatrix}$$

which is invertible for any (x, w) in $\mathbb{R}^{13}$ and any time t. In fact, as for the high-gain observer for the oscillator, $T_e(\cdot, t)$ is a diffeomorphism on $\mathbb{R}^{13}$ such that $T_e(\mathbb{R}^{13}, t) =$

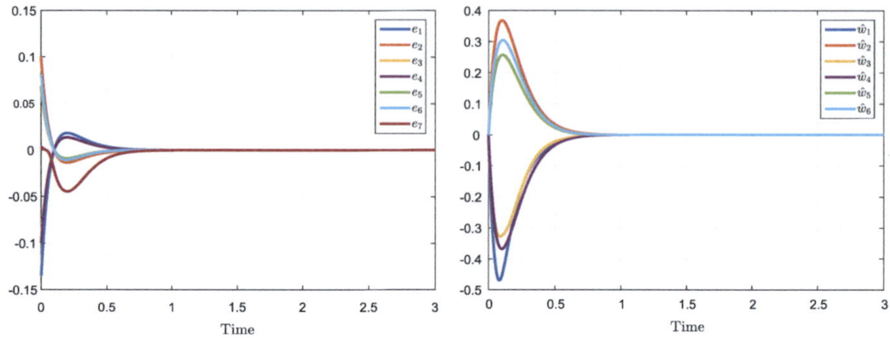

(a) $e_i = \hat{x}_i - x_i$ (b) $\hat{w}_i$

Fig. 11.3 Observer (11.23) with $L = 10$ and $k_1 = k_2 = 1$

$\mathbb{R}^{13}$ for any t. Thus, we have managed to transform an injective immersion $T(\cdot, t)$: $\mathbb{R}^7 \to \mathbb{R}^{13}$ into a surjective diffeomorphism $T_e(\cdot, \cdot, t) : \mathbb{R}^{13} \to \mathbb{R}^{13}$.

Observer in the given coordinates As suggested at the beginning of this section, we consider the observer:

$$\dot{\overbrace{\begin{bmatrix} \hat{x} \\ \hat{w} \end{bmatrix}}} = \left(\frac{\partial T_e}{\partial(x,w)}(\hat{x},\hat{w},t) \right)^{-1} \left(\varphi(T_e(\hat{x},\hat{w},t),\hat{x},t,y) - \frac{\partial T_e}{\partial t}(\hat{x},\hat{w},t) \right) \tag{11.23}$$

where φ is defined on $\mathbb{R}^{13} \times \mathbb{R}^7 \times \mathbb{R} \times \mathbb{R}^6$ by

$$\varphi(\hat{\xi}, \hat{x}, t, y) = \begin{pmatrix} \hat{\xi}_d + Lk_1(y - \hat{\xi}_m) \\ \mathsf{sat}(L_f^2 h(\hat{x}, u(t), \dot{u}(t)), \overline{\mathsf{fl}}) + L^2 k_2(y - \hat{\xi}_m) \\ -a\hat{\xi}_a \end{pmatrix}$$

with $\hat{\xi} = (\hat{\xi}_m, \hat{\xi}_d, \hat{\xi}_a) \in \mathbb{R}^6 \times \mathbb{R}^6 \times \mathbb{R}$, a any strictly positive number. A result of a simulation is given in Fig. 11.3.

References

1. Andrieu, V., Eytard, J.B., Praly, L.: Dynamic extension without inversion for observers. In: IEEE Conference on Decision and Control, pp. 878–883 (2014)
2. Bernard, P., Praly, L., Andrieu, V.: Expressing an observer in preferred coordinates by transforming an injective immersion into a surjective diffeomorphism. SIAM J. Control Optim. **56**(3), 2327–2352 (2018)
3. Gibert, V.: Analyse d'observabilité et synthèse d'observateurs robustes pour l'atterrissage basé vision d'avions de ligne sur des pistes inconnues. Ph.D. thesis, Ecole Centrale de Nantes (2016)
4. Gibert, V., Burlion, L., Chriette, A., Boada, J., Plestan, F.: New pose estimation scheme in perspective vision system during civil aircraft landing. In: IFAC Symposium on Robot Control (2015)

5. Gibert, V., Burlion, L., Chriette, A., Boada, J., Plestan, F.: Nonlinear observers in vision system: application to civil aircraft landing. In: European Control Conference (2015)
6. Praly, L., Marconi, L., Isidori, A.: A new observer for an unknown harmonic oscillator. In: Symposium on Mathematical Theory of Networks and Systems (2006)

Appendix A
Technical Lemmas

In this appendix, we give the proof to some general technical lemmas used throughout this book.

A.1 About Homogeneity

Lemma A.1 *Let η be a continuous functions defined on $\mathbb{R}^{n+1}$ and f a continuous function defined on $\mathbb{R}^n$. Let $\mathscr{C}$ be a compact subset of $\mathbb{R}^n$. Assume that, for all x in $\mathscr{C}$ and s in $\mathrm{S}(f(x))$,*

$$\eta(x, s) < 0 \, .$$

Then, there exists $\alpha > 0$ such that for all x in $\mathscr{C}$ and s in $\mathrm{S}(f(x))$

$$\eta(x, s) < -\alpha \, .$$

Proof Assume that for all $k > 0$, there exists x_k in $\mathscr{C}$ and s_k in $\mathrm{S}(f(x_k)) \subset [-1, 1]$ such that

$$0 > \eta(x_k, s_k) \geq -\frac{1}{k} \, .$$

Then, $\eta(x_k, s_k)$ tends to 0 when k tends to infinity. Besides, there exists a subsequence (k_m) such that x_{k_m} tends to x^* in $\mathscr{C}$ and s_{k_m} tends to s^* in $[-1, 1]$. Since η is continuous, it follows that $\eta(x^*, s^*) = 0$ and we will have a contradiction if $s^* \in \mathrm{S}(f(x^*))$. If $f(x^*)$ is not zero, then by continuity of f, s^* is equal to the sign of $f(x^*)$, and otherwise, $s^* \in [-1, 1] = \mathrm{S}(f(x^*))$. Thus, $s^* \in \mathrm{S}(f(x^*))$ in all cases. □

Lemma A.2 *Let η be a function defined on $\mathbb{R}^n$ homogeneous with degree d and weight vector $r = (r_1, ..., r_n)$, and V a positive definite proper function defined on*

P. Bernard, *Observer Design for Nonlinear Systems*,
Lecture Notes in Control and Information Sciences 479,
https://doi.org/10.1007/978-3-030-11146-5

$\mathbb{R}^n$ homogeneous of degree d_V with same weight vector r. Define $\mathscr{C} = V^{-1}(\{1\})$. If there exists α such that for all x in $\mathscr{C}$

$$\eta(x) < \alpha ,$$

then for all x in $\mathbb{R}^n \setminus \{0\}$,

$$\eta(x) < \alpha V(x)^{\frac{d}{d_V}} .$$

Proof Let x in $\mathbb{R}^n \setminus \{0\}$. We have $\bar{x} = \dfrac{x_i}{V(x)^{\frac{r_i}{d_V}}}$ in $\mathscr{C}$. Thus, $\eta(\bar{x}) < \alpha$ and by homogeneity

$$\frac{1}{V(x)^{\frac{d}{d_V}}}\eta(x) < \alpha$$

which gives the required inequality. □

Lemma A.3 *Let η be a homogeneous function of degree d and weight vector r defined on $\mathbb{R}^n$ by*

$$\eta(x) = \max_{s \in S(f(x))} \tilde{\eta}(x, s)$$

where $\tilde{\eta}$ is a continuous function defined on $\mathbb{R}^{n+1}$ and f a continuous function defined on $\mathbb{R}^n$. Consider a continuous function γ homogeneous with same degree and weight vector such that, for all x in $\mathbb{R}^n \setminus \{0\}$ and s in $S(f(x))$

$$\begin{aligned} &\gamma(x) \geq 0 , \\ &\gamma(x) = 0 \quad \Rightarrow \quad \tilde{\eta}(x, s) < 0 . \end{aligned}$$

Then, there exists $k_0 > 0$ such that, for all x in $\mathbb{R}^n \setminus \{0\}$,

$$\eta(x) - k_0\, \gamma(x) < 0 .$$

Proof Define the homogeneous definite positive function $V(x) = \sum_{i=1}^{n} |x_i|^{\frac{d}{r_i}}$, and consider the compact set $\mathscr{C} = V^{-1}(\{1\})$. Assume that for all $k > 0$, there exists x_k in $\mathscr{C}$ and s_k in $S(f(x_k))$ such that

$$\tilde{\eta}(x_k, s_k) \geq k\, \gamma(x_k) \geq 0$$

$\tilde{\eta}$ is continuous and thus bounded on the compact set $\mathscr{C} \times [-1, 1]$. Therefore, $\gamma(x_k)$ tends to 0 when k tends to infinity. Besides, there exists a subsequence (k_m) such that x_{k_m} tends to x^* in $\mathscr{C}$ and s_{k_m} tends to s^* in $[-1, 1]$. It follows that $\gamma(x^*) = 0$ since γ is continuous. But with the same argument as in the proof of Lemma A.1, we have $s^* \in \mathtt{S}(f(x^*))$. It yields that $\tilde{\eta}(x^*, s^*) < 0$ by assumption, and we have a contradiction.

Therefore, there exists k_0 such that

$$\tilde{\eta}(x, s) - k_0\, \gamma(x) < 0$$

for all x in $\mathscr{C}$ and all s in $\mathtt{S}(f(x))$. Thus, with Lemma A.1 there exists $\alpha > 0$ such that

$$\tilde{\eta}(x, s) - k_0\, \gamma(x) \le -\alpha$$

so that

$$\eta(x) - k_0\, \gamma(x) < 0$$

for any x in $\mathscr{C}$. The result follows applying Lemma A.2. □

Lemma A.4 *Consider a positive bounded continuous function $t \mapsto c(t)$ and an absolutely continuous function $t \mapsto \nu(t)$ both defined on $[0, \bar{t})$ and such that for almost all t in $[0, \bar{t})$ such that $\nu(t) \ge c(t)$ then $\dot{\nu}(t) \le -\nu(t)^d$ with d in $]0, 1[$. Then, for all t in $[0, \bar{t})$,*

$$\nu(t) \le \max\left\{0, \max\{\nu(0) - c(0), 0\}^{1-d} - t\right\}^{1/(1-d)} + \sup_{s \in [0,t]} c(s)\ .$$

Proof Let t be in $[0, \bar{t})$ and $c_t = \sup_{s \in [0,t]} c(s)$. For almost all $s \le t$ such that $\nu(s) \ge \nu_t$, $\dot{\nu}(s) \le -\nu(s)^d$, and thus

$$\begin{aligned} \overbrace{\max\{\nu(s) - c_t, 0\}}^{\cdot} &\le -\nu(s)^d \\ &\le -\max\{\nu(s) - c_t, 0\}^d\ . \end{aligned}$$

This inequality is also true when $\nu(s) < c_t$; therefore, it is true for almost all $s \le t$. It follows that for all $s \le t$

$$\begin{aligned} \max\{\nu(s) - c_t, 0\}^{1-d} &\le \max\{\nu(0) - c_t, 0\}^{1-d} - s \\ &\le \max\{\nu(0) - c(0), 0\}^{1-d} - s\ , \end{aligned}$$

i.e.,

$$\max\{\nu(s) - c_t, 0\} \le \max\left\{0, \left\{\max\{\nu(0) - c(0), 0\}^{1-d} - s\right\}\right\}^{\frac{1}{1-d}}$$

and finally, for all $s \le t$

$$\nu(s) \le \max\left\{0, \left\{\max\{\nu(0) - c(0), 0\}^{1-d} - s\right\}\right\}^{\frac{1}{1-d}} + c_t.$$

Taking this inequality at $s = t$ gives the required result. □

Lemma A.5 *For any* (x_a, x_b) *in* $\mathbb{R}^2$*, for any* $p \geq 1$*, we have*

- $\left| \lfloor x_a \rceil^{\frac{1}{p}} - \lfloor x_b \rceil^{\frac{1}{p}} \right| \leq 2^{1-\frac{1}{p}} \left| x_a - x_b \right|^{\frac{1}{p}}$
- $(|x_a| + |x_b|)^{\frac{1}{p}} \leq |x_a|^{\frac{1}{p}} + |x_b|^{\frac{1}{p}}$.

Proof The second inequality is just the definition of the concavity of $x \mapsto x^{\frac{1}{p}}$ on $\mathbb{R}^+$. As for the first one, it is enough to prove it for $|x_a| \geq |x_b|$ (otherwise exchange them) and x_a nonnegative (otherwise take $(-x_a, -x_b)$). Besides, since it clearly holds for $x_b = 0$, we only have to prove (for $x = \frac{x_a}{|x_b|}$),

$$x^{\frac{1}{p}} \pm 1 \leq 2^{1-\frac{1}{p}} (x \pm 1)^{\frac{1}{p}} \quad \forall x \geq 1 \, .$$

First, by concavity of $x \mapsto x^{\frac{1}{p}}$, $\frac{1}{2} x^{\frac{1}{p}} + \frac{1}{2} 1^{\frac{1}{p}} \leq \left(\frac{x+1}{2}\right)^{\frac{1}{p}}$ which gives the required inequality for the case "+". Besides, still by concavity of $x \mapsto x^{\frac{1}{p}}$, we have for $x \geq 1$, $\frac{x-1}{x} x^{\frac{1}{p}} + \frac{1}{x} 0^{\frac{1}{p}} \leq \left(\frac{x-1}{x} x + \frac{1}{x} 0\right)^{\frac{1}{p}}$ and $\frac{1}{x} x^{\frac{1}{p}} + \frac{x-1}{x} 0^{\frac{1}{p}} \leq \left(\frac{1}{x} x + \frac{x-1}{x} 0\right)^{\frac{1}{p}}$. Adding those two inequalities gives the case "−". □

A.2 About Continuity

Lemma A.6 *Let* $\psi : \mathbb{R}^n \to \mathbb{R}^q$ *be a continuous function on a compact subset* $\mathcal{C}$ *of* $\mathbb{R}^n$*. There exists a concave class* $\mathcal{K}$ *function* ρ *such that for all* (x_a, x_b) *in* $\mathcal{C}^2$

$$|\psi(x_a) - \psi(x_b)| \leq \rho(|x_a - x_b|) \, .$$

Proof Define the function

$$\rho_0(s) = \max_{x \in \mathcal{C}, |e| \leq s} |\psi(x + e) - \psi(x)|$$

which is increasing and such that $\rho_0(0) = 0$. Let us show that it is continuous at 0. Let (s_n) a sequence converging to 0. For all n, there exists x_n in $\mathcal{C}$ and e_n such that $|e_n| \leq s_n$ and $\rho_0(s_n) = |\psi(x_n + e_n) - \psi(x_n)|$. Since $\mathcal{C}$ is compact, there exist x^* in $\mathcal{C}$, e^* and subsequences of (x_n) and (e_n) converging to x^* and e^*, respectively. But e^* is necessarily 0 and by continuity of ψ, $\rho_0(s_n)$ tends to 0.

Now, the function, defined by the Riemann integral

$$\rho_1(s) = \begin{cases} \frac{1}{s} \displaystyle\int_s^{2s} \rho_0(s) ds + s \, , & s > 0 \\ 0 & , \, s = 0 \end{cases}$$

is continuous, strictly increasing and such that $\rho_0(s) \leq \rho_1(s)$. Besides, taking $\bar{s} = \max_{(x_a, x_b) \in \mathcal{C}^2} |x_a - x_b|$, there exists a concave class $\mathcal{K}$ function ρ such that for all s

in $[0, \bar{s}]$, $\rho_1(s) \leq \rho(s)$ (see [2] for instance). Finally, we have:

$$|\psi(x_a) - \psi(x_b)| \leq \rho(|x_a - x_b|) \qquad \forall (x_a, x_b) \in \mathscr{C}^2 .$$

□

Lemma A.7 *Consider a function $\psi : \mathbb{R}^n \to \mathbb{R}$. Assume that there exist a compact set $\mathscr{C}$ of $\mathbb{R}^n$ and a function ρ of class $\mathscr{K}$ such that for all (x_a, x_b) in $\mathscr{C}^2$*

$$|\psi(x_a) - \psi(x_b)| \leq \rho(|x_a - x_b|) .$$

Define the function $\hat{\psi} : \mathbb{R}^n \to [-\overline{\psi}, \overline{\psi}]$ by[1]

$$\hat{\psi}(z) = \mathsf{sat}_{\overline{\psi}}(\psi(z))$$

with $\overline{\psi} = \max_{z \in \mathscr{C}} \psi(z)$. Then, for any compact subset $\tilde{\mathscr{C}}$ strictly contained[2] *in $\mathscr{C}$, there exists a positive real number c such that for all (x_a, x_b) in $\mathbb{R}^n \times \tilde{\mathscr{C}}$,*

$$|\hat{\psi}(x_a) - \hat{\psi}(x_b)| \leq c\rho(|x_a - x_b|) \tag{A.1}$$

Proof Since $\mathscr{C}$ strictly contains $\tilde{\mathscr{C}}$, we have:

$$\delta = \inf_{(x_a, x_b) \in (\mathbb{R}^n \setminus \mathscr{C}) \times \tilde{\mathscr{C}}} |x_a - x_b| > 0 .$$

First, for x_b in $\tilde{\mathscr{C}}$, $\hat{\psi}(x_b) = \psi(x_b)$. Now, if x_a is in $\mathscr{C}$, then we have $\hat{\psi}(x_a) = \psi(x_a)$, and consequently, (A.1) holds for $c \geq 1$. If $x_a \notin \mathscr{C}$, we have, for all x_b in $\tilde{\mathscr{C}}$,

$$\begin{gathered} |x_a - x_b| \geq \delta , \\ |\hat{\psi}(x_a) - \hat{\psi}(x_b)| \leq 2\overline{\psi} \leq 2\overline{\psi}\frac{\rho(|x_a - x_b|)}{\rho(\delta)} , \end{gathered}$$

and (A.1) holds for $c \geq \frac{2\overline{\psi}}{\rho(\delta)}$. □

A.3 About Injectivity

In this appendix, we consider two continuous functions $\Psi : \mathbb{R}^n \to \mathbb{R}^r$ and $\gamma : \mathbb{R}^n \to \mathbb{R}^q$ and a subset $\mathscr{S}$ of $\mathbb{R}^n$ such that

$$\Psi(x_a) = \Psi(x_b) \quad \forall (x_a, x_b) \in \mathscr{S}^2 : \gamma(x_a) = \gamma(x_b) . \tag{A.2}$$

---

[1] The saturation function $\mathsf{sat}_M(\cdot)$ is defined by $\mathsf{sat}_M(x) = \max\{\min\{x, M\}, -M\}$.

[2] By strictly contained, we mean that $\tilde{C} \subset \mathscr{C}$ and the distance between $\tilde{\mathscr{C}}$ and the complement of $\mathscr{C}$, namely $\mathbb{R}^n \setminus \mathscr{C}$, is strictly positive.

In the particular case where Ψ is the identity function, (A.2) characterizes the injectivity of γ.

Lemma A.8 *There exists a function* ψ *defined on* $\gamma(\mathscr{S})$ *such that*

$$\Psi(x) = \psi(\gamma(x)) \quad \forall x \in \mathscr{S} \,. \tag{A.3}$$

Proof Define the map ψ on $\gamma(\mathscr{S})$ as

$$\psi(z) = \bigcup_{\substack{x \in \mathscr{S} \\ \gamma(x)=z}} \{\Psi(x)\} \,.$$

For any z in $\gamma(\mathscr{S})$, the set $\psi(z)$ is non-empty and single-valued because according to (A.2), if $z = \gamma(x_a) = \gamma(x_b)$, then $\Psi(x_a) = \Psi(x_b)$. Therefore, we can consider ψ as a function defined on $\gamma(\mathscr{S})$ and it verifies (A.3). □

Lemma A.9 *Consider any compact subset* $\mathscr{C}$ *of* $\mathscr{S}$. *There exists a concave class* $\mathscr{K}$ *function* ρ *such that for all* (x_a, x_b) *in* $\mathscr{C}^2$

$$|\Psi(x_a) - \Psi(x_b)| \leq \rho\left(|\gamma(x_a) - \gamma(x_b)|\right) \,. \tag{A.4}$$

Proof We denote $D(x_a, x_b) = |\gamma(x_a) - \gamma(x_b)|$. Let

$$\rho_0(s) = \max_{\substack{(x_a, x_b) \in \mathscr{C}^2 \\ D(x_a, x_b) \leq s}} |\Psi(x_a) - \Psi(x_b)|$$

This defines properly a non-decreasing function with nonnegative values which satisfies:

$$|\Psi(x_a) - \Psi(x_b)| \;\leq\; \rho_0(D(x_a, x_b)) \qquad \forall (x_a, x_b) \in \mathscr{C}^2 \,.$$

Also $\rho_0(0) = 0$. Indeed if not there would exist (x_a, x_b) in $\mathscr{C}^2$ satisfying:

$$D(x_a, x_b) \;=\; 0 \quad , \qquad |\Psi(x_a) - \Psi(x_b)| \;>\; 0 \,.$$

But this contradicts Eq. (A.2).

Moreover, it can be shown that this function is also continuous at $s = 0$. Indeed, let $(s_k)_{k\in\mathbb{N}}$ be a sequence converging to 0. For each k, there exist $(x_{a,k}, x_{b,k})$ in $\mathscr{C}^2$ which satisfies $D(x_{a,k}, x_{b,k}) \leq s_k$ and $\rho_0(s_k) = |\Psi(x_{a,k}) - \Psi(x_{b,k})|$. The sequence $(x_{a,k}, x_{b,k})_{k\in\mathbb{N}}$ being in a compact set, it admits an accumulation point (x_a^*, x_b^*) which, because of the continuity of D, must satisfy $D(x_a^*, x_b^*) = 0$ and therefore with (A.2) also $\Psi(x_a^*) - \Psi(x_b^*) \;=\; 0$. It follows that $\rho_0(s_k)$ tends to 0 and ρ_0 is continuous at 0. Proceeding with the same regularization of ρ_0 as in the proof of Lemma A.6, the conclusion follows. □

Lemma A.10 *Consider any compact subset $\mathscr{C}$ of $\mathscr{S}$. There exists a uniformly continuous function ψ defined on $\mathbb{R}^q$ such that*

$$\Psi(x) = \psi(\gamma(x)) \quad \forall x \in \mathscr{C} .$$

Proof Consider ψ and ρ given by Lemmas A.8 and A.9, respectively. For any (z_a, z_b) in $\gamma(\mathscr{C})^2$, there exists (x_a, x_b) in $\mathscr{C}^2$ such that $z_a = \gamma(x_a)$ and $z_b = \gamma(x_b)$. Applying (A.4) to (x_a, x_b) and using (A.3), we have

$$|\psi(z_a) - \psi(z_b)| \;\leq\; \rho(|z_a - z_b|) .$$

ρ being concave, we deduce from [2, Theorem 2] (applied to each of the r real-valued components of ψ) that ψ admits a uniformly continuous extension defined on $\mathbb{R}^q$. Note that the extension of each component preserves the modulus of continuity ρ, so that the global extension has a modulus of continuity equal to $c\rho$ for some $c > 0$ depending only on the choice of the norm on $\mathbb{R}^r$. □

When $q \leq n$ and γ is full-rank on $\mathscr{C}$, the function ψ is even C^1:

Lemma A.11 *Assume that $q \leq n$ and $\frac{\partial \gamma}{\partial x}$ is full-rank on $\mathscr{S}$, namely γ is a submersion on $\mathscr{S}$. Then, $\gamma(\mathscr{S})$ is open and there exists a C^1 function ψ defined on $\gamma(\mathscr{S})$ such that*

$$\Psi(x) = \psi(\gamma(x)) \quad \forall x \in \mathscr{S} .$$

Proof γ is an open map according to [1, Proposition 4.28], and thus, $\gamma(\mathscr{S})$ is open. Consider the function ψ given by Lemma A.8, and take any z^* in $\gamma(\mathscr{S})$. There exists x^* in $\mathscr{S}$ such that $z^* = \gamma(x^*)$. γ being full-rank at x^*, according to the constant rank theorem, there exists an open neighborhood $\mathscr{V}$ of x^* and C^1 diffeomorphisms $\psi_1 : \mathbb{R}^n \to \mathscr{V}$ and $\psi_2 : \mathbb{R}^q \to \gamma(\mathscr{V})$ such that for all $\tilde{x}$ in $\mathbb{R}^n$:

$$\gamma(\psi_1(\tilde{x})) = \psi_2(\tilde{x}_1, \ldots, \tilde{x}_q) .$$

It follows that for all z in $\gamma(\mathscr{V})$

$$\gamma(\psi_1(\psi_2^{-1}(z), 0)) = z$$

namely γ admits a C^1 right-inverse γ^{ri} defined on $\gamma(\mathscr{V})$ which is an open neighborhood of z^*. Therefore, $\psi = \Psi \circ \gamma^{ri}$ and ψ is C^1 at z^*. □

A direct consequence from those results is that any continuous function $\gamma : \mathbb{R}^n \to \mathbb{R}^q$ injective on a compact set $\mathscr{C}$ admits a uniformly continuous left inverse ψ defined on $\mathbb{R}^q$ (take $\Psi = \mathsf{Id}$). The previous lemma does not apply because γ cannot be a submersion. However, we will show now that when γ is full-rank (i.e., an immersion), this left inverse can be taken Lipschitz on $\mathbb{R}^q$.

Due to needs in Chaps. 6 and 7, we generalize those results to the case where the function γ depends on another parameter w evolving in a compact set:

Lemma A.12 *Let $\gamma : \mathbb{R}^n \times \mathbb{R}^p \to \mathbb{R}^q$ be a continuous function and compact sets $\mathscr{C}_x$ and $\mathscr{C}_w$ of $\mathbb{R}^n$ and $\mathbb{R}^p$, respectively, such that for all w in $\mathscr{C}_w$, $x \mapsto \gamma(x, w)$ is injective on $\mathscr{C}_x$.*

Then, there exists a concave class $\mathscr{K}$ function ρ, such that for all (x_a, x_b) in $\mathscr{C}_x^2$ and all w in $\mathscr{C}_w$,

$$|x_a - x_b| \leq \rho(|\gamma(x_a, w) - \gamma(x_b, w)|) ,$$

and a function ψ defined on $\mathbb{R}^q \times \mathbb{R}^p$ and a strictly positive number c such that

$$x = \psi(\gamma(x, w), w) \quad \forall (x, w) \in \mathscr{C}_x \times \mathscr{C}_w$$

and

$$|\psi(z_a, w) - \psi(z_b, w)| \leq c\rho(|z_a - z_b|)$$

That is, $z \mapsto \psi(z, w)$ is uniformly continuous on $\mathbb{R}^q$, uniformly in w.

If besides for all w in $\mathscr{C}_w$, $x \mapsto \gamma(x, w)$ is an immersion on $\mathscr{C}_x$, i.e., for all w in $\mathscr{C}_w$, and all x in $\mathscr{C}_x$, $\dfrac{\partial \gamma}{\partial x}(x, w)$ is full-rank, then ρ is linear and $z \mapsto \psi(z, w)$ is Lipschitz on $\mathbb{R}^q$, uniformly in w.

Proof The proof of the existence of ρ follows exactly that of Lemma A.9, but adding in the max defining ρ_0, $w \in \mathscr{C}_w$. Since it is a compact set, ρ is well defined and the same ρ can then be used for any w in $\mathscr{C}_w$. Applying Lemma A.10 to every $x \mapsto \gamma(x, w)$ gives the result since it is shown there that the extensions admit all the same modulus of continuity $c\rho$ for some $c > 0$ depending only on the norm chosen on $\mathbb{R}^r$.

Now suppose that $x \mapsto \gamma(x, w)$ is full-rank for all w in $\mathscr{C}_w$. Let Δ be the function defined on $\mathscr{C}_x \times \mathscr{C}_x \times \mathscr{C}_w$ by

$$\Delta(x_a, x_b, w) = \gamma(x_a, w) - \gamma(x_b, w) - \frac{\partial \gamma}{\partial x}(x_b, w)(x_a - x_b) .$$

Since $\frac{\partial \gamma}{\partial x}(x, w)$ is full-rank by assumption, the function

$$P(x, w) = \left(\frac{\partial \gamma}{\partial x}(x, w)^\top \frac{\partial \gamma}{\partial x}(x, w) \right)^{-1} \frac{\partial \gamma}{\partial x}(x, w)^\top$$

is well-defined and continuous on $\mathscr{C}_x \times \mathscr{C}_w$, and for any (x_a, x_b, w) in $\mathscr{C}_x \times \mathscr{C}_x \times \mathscr{C}_w$, we have

$$|x_a - x_b| \leq P_m(|\gamma(x_a, w) - \gamma(x_b, w| + |\Delta(x_a, x_b, w)|)$$

with $P_m = \max_{\mathscr{C}_x \times \mathscr{C}_w} |P(x, w)|$. Besides, the function $\frac{|\Delta(x_a, x_b, w)|}{|x_a - x_b|^2}$ is defined and continuous on $\mathscr{C}_x \times \mathscr{C}_x \times \mathscr{C}_w$; thus, there exists $L_\Delta > 0$ such that

$$|\Delta(x_a, x_b, w)| \leq L_\Delta |x_a - x_b|^2 \leq \frac{1}{2P_m}|x_a - x_b|$$

for any (x_a, x_b) in $\mathscr{C}_x^2$ such that $|x_a - x_b| \leq 2r$ with $r = \frac{1}{4P_m L_\Delta}$, and for any w in $\mathscr{C}_w$. Now, define the set

$$\Omega = \{(x_a, x_b) \in \mathscr{C}_x^2 \mid |x_a - x_b| \geq 2r\}$$

which is a closed subset of the compact set $\mathscr{C}_x^2$ and therefore compact. The function $(x_a, x_b, w) \mapsto \frac{|x_a - x_b|}{|\gamma(x_a, w) - \gamma(x_b, w)|}$ is defined and continuous on $\Omega \times \mathscr{C}_w$ since $\gamma(\cdot, w)$ is injective for any w in $\mathscr{C}_w$. Thus, it admits a maximum M on the compact set $\Omega \times \mathscr{C}_w$.

Finally, take any (x_a, x_b) in $\mathscr{C}_x^2$ and any w in $\mathscr{C}_w$. There are two cases:

- either $(x_a, x_b) \notin \Omega$, i.e., $|x_a - x_b| < 2r$, and

$$|x_a - x_b| \leq \frac{P_m}{2} |\gamma(x_a, w) - \gamma(x_b, w)| .$$

- or $(x_a, x_b) \in \Omega$, and

$$|x_a - x_b| \leq M |\gamma(x_a, w) - \gamma(x_b, w)| .$$

We conclude that ρ can be chosen linear with rate $L = \max\{\frac{P_m}{2}, M\}$. □

References

1. Lee, J.M.: Introduction to Smooth Manifolds. Springer, Berlin (2013)
2. McShane, E.J.: Extension of range of functions. Bull. Am. Math. Soc. **40**(12), 837–842 (1934)

Appendix B
Lyapunov Analysis for High-Gain Homogeneous Observers

In this appendix, we give the Lyapunov-based proofs of convergence of the main high-gain type observers presented in Chap. 4 for triangular normal forms. They all consist in showing that the triangular nonlinearities can be dominated by taking a sufficiently large gain.

B.1 Standard High-Gain Observer

B.1.1 ... For a Lipschitz Triangular Form

We prove here the asymptotic convergence of the high-gain observer (4.4) for the Lipschitz triangular form (4.2), namely Theorem 4.1.

Proof The error $e = \hat{\xi} - \xi$ produced by the observer (4.4) satisfies

$$\dot{e} = LAe + \Phi(u, \hat{\xi}) - \Phi(u, \xi) - L\mathscr{L}KCe \tag{B.1}$$

where

$$A = \begin{pmatrix} 0 & I_{d_y} & 0 & \cdots & 0 \\ \vdots & \ddots & \ddots & & \vdots \\ \vdots & & \ddots & \ddots & 0 \\ & & & 0 & I_{d_y} \\ 0 & \cdots & \cdots & 0 & 0 \end{pmatrix} \quad , \quad \mathscr{L} = \begin{pmatrix} I_{d_y} & & & \\ & LI_{d_y} & & \\ & & \ddots & \\ & & & L^{m-1}I_{d_y} \end{pmatrix}$$

$$C = \begin{pmatrix} I_{d_y} & 0 \cdots 0 \end{pmatrix} \quad , \quad K = \begin{pmatrix} k_1 I_{d_y} \\ k_2 I_{d_y} \\ \vdots \\ k_m I_{d_y} \end{pmatrix} .$$

P. Bernard, *Observer Design for Nonlinear Systems*,
Lecture Notes in Control and Information Sciences 479,
https://doi.org/10.1007/978-3-030-11146-5

Consider the scaled error coordinates $\varepsilon = \mathscr{L}^{-1}e$. It is straightforward to see that in those coordinates, those error dynamics read

$$\frac{1}{L}\dot{\varepsilon} = (A - KC)\varepsilon + \frac{1}{L}\mathscr{L}^{-1}\left(\Phi(u, \hat{\xi}) - \Phi(u, \xi)\right) \tag{B.2}$$

Because of the choice of the k_i, the matrix $A - KC$ is Hurwitz, so there exists a positive definite matrix P and a strictly positive real number λ such that

$$(A - KC)^\top P + P(A - KC) \leq -4\lambda P \ .$$

Define the function

$$V(\varepsilon) = \varepsilon^\top P \varepsilon$$

which is positive definite and radially unbounded and verifies

$$\alpha_1|\varepsilon|^2 \leq V(\varepsilon) \leq \alpha_2|\varepsilon|^2 \qquad \forall \varepsilon \in \mathbb{R}^{d_\xi}$$

for some positive real numbers α_1 and α_2 (the smallest and largest eigenvalue of P, respectively). Computing the derivative of V along the trajectories, we get

$$\frac{1}{L}\dot{\overline{V(\varepsilon)}} = -4\lambda V(\varepsilon) + \frac{2}{L}\varepsilon^\top P\mathscr{L}^{-1}\left(\Phi(u, \hat{\xi}) - \Phi(u, \xi)\right) \tag{B.3}$$

But since Φ satisfies $\mathscr{H}(\alpha, \mathfrak{a})$, we have[3] for all $L \geq 1$,

$$\begin{aligned}\left|\frac{1}{L^{i-1}}\left(\Phi_i(u, \hat{\boldsymbol{\xi}}_i) - \Phi_i(u, \boldsymbol{\xi}_i)\right)\right| &\leq \mathfrak{a}\sum_{j=1}^{i}\frac{1}{L^{i-1}}|e_j| \\ &\leq \mathfrak{a}\sum_{j=1}^{i}\frac{1}{L^{j-1}}|e_j| = \mathfrak{a}\sum_{j=1}^{i}|\varepsilon_j| \\ &\leq \mathfrak{a}|\varepsilon|_1 \leq \mathfrak{a}\, d_\xi|\varepsilon|\end{aligned} \tag{B.4}$$

and therefore,

$$\left|\mathscr{L}^{-1}\left(\Phi(u, \hat{\xi}) - \Phi(u, \xi)\right)\right| \leq m\, \mathfrak{a}\, d_\xi|\varepsilon|$$

so that[4]

---

[3]We denote $|\varepsilon|_1$ the 1-norm, while $|\varepsilon|$ denotes the 2-norm.

[4]Using the Cauchy–Schwartz inequality in the norm associated with P: $x^\top Py \leq \sqrt{x^\top Px}\sqrt{y^\top Py}$.

$$\varepsilon^\top P \mathscr{L}^{-1}\left(\Phi(u,\hat{\xi}) - \Phi(u,\xi)\right) \leq \sqrt{\varepsilon^\top P \varepsilon}\,\sqrt{\alpha_2}\left|\mathscr{L}^{-1}\left(\Phi(u,\hat{\xi}) - \Phi(u,\xi)\right)\right|$$
$$\leq m\, d_\xi \sqrt{\frac{\alpha_2}{\alpha_1}}\,\mathfrak{a}\, V(\varepsilon)\,.$$

We conclude that

$$\frac{1}{L}\,\dot{\overline{V(\varepsilon)}} \leq -2\left(2\lambda - m\, d_\xi \sqrt{\frac{\alpha_2}{\alpha_1}}\frac{\mathfrak{a}}{L}\right) V(\varepsilon)$$

and thus for any L verifying

$$L \geq \max\{\mathfrak{a}\, L^*, 1\} \quad , \quad L^* = \frac{m\, d_\xi}{\lambda}\sqrt{\frac{\alpha_2}{\alpha_1}}\,,$$

we have

$$V(\varepsilon(t)) \leq V(\varepsilon(t_0))\, e^{-2\lambda L(t-t_0)}\,,$$

which gives

$$|\varepsilon(t)| \leq \sqrt{\frac{\alpha_2}{\alpha_1}}\,|\varepsilon(t_0)|\, e^{-\lambda L(t-t_0)}\,.$$

The result follows from the fact that $e_i = L^{i-1}\varepsilon_i$. □

B.1.2 ... For a Hölder Triangular Form

We prove here the practical convergence of the high-gain observer (4.4) for the triangular form (4.2) verifying the Hölder-type conditions (4.5), namely Theorem 4.2 as in [2].

Proof With Young's inequality, we obtain from (4.3) that, for all σ_{ij} in $\mathbb{R}_+$ and all $\hat{\xi}$ and ξ in $\mathbb{R}^{d_\xi}$

$$\left|\Phi_i(u,\hat{\boldsymbol{\xi}}_i) - \Phi_i(u,\boldsymbol{\xi}_i)\right| \;\leq\; \sum_{j=1}^{i} \mathfrak{a}_{ij}|\hat{\xi}_j - \xi_j| + \mathfrak{b}_{ij}\,, \tag{B.5}$$

with $\mathfrak{a}_{ij}$ and $\mathfrak{b}_{ij}$ defined as

$$\begin{cases} \mathfrak{a}_{ij} = 0\,,\ \mathfrak{b}_{ij} = \mathfrak{a}\,, & \text{if } \alpha_{ij} = 0 \\ \mathfrak{a}_{ij} = \mathfrak{a}^{\frac{1}{\alpha_{ij}}}\alpha_{ij}\sigma_{ij}^{\frac{1}{\alpha_{ij}}}\,,\ \mathfrak{b}_{ij} = \dfrac{1-\alpha_{ij}}{\sigma_{ij}^{\frac{1}{1-\alpha_{ij}}}} & \text{if } 0 < \alpha_{ij} < 1 \\ \mathfrak{a}_{ij} = \mathfrak{a}\,,\ \mathfrak{b}_{ij} = 0 & \text{if } \alpha_{ij} = 1 \end{cases} \tag{B.6}$$

It follows that for all $L \geq 1$,

$$
\begin{aligned}
\left|\frac{1}{L^{i-1}}\left(\Phi_i(u,\hat{\xi}_i) - \Phi_i(u,\xi_i)\right)\right| &\leq \sum_{j=1}^{i} \frac{1}{L^{i-1}} \mathfrak{a}_{ij}|e_j| + \frac{1}{L^{i-1}}\mathfrak{b}_{ij} \\
&\leq \tilde{\mathfrak{a}} \sum_{j=1}^{i} |\varepsilon_j| + \frac{1}{L^{i-1}}\mathfrak{b}_i \\
&\leq \tilde{\mathfrak{a}}\, d_\xi\, |\varepsilon| + \frac{1}{L^{i-1}}\mathfrak{b}_i
\end{aligned}
\tag{B.7}
$$

with

$$
\tilde{\mathfrak{a}} = \max_{i \geq j} \mathfrak{a}_{ij} \quad , \quad \mathfrak{b}_i = \sum_{j=1}^{i} \mathfrak{b}_{ij} \,. \tag{B.8}
$$

Reproducing the proof of Theorem 4.1 with (B.7) instead of (B.4),

$$
\left|\mathscr{L}^{-1}\left(\Phi(u,\hat{\xi}) - \Phi(u,\xi)\right)\right| \leq \frac{m d_\xi}{\sqrt{\alpha_1}}\, \tilde{\mathfrak{a}} \sqrt{\varepsilon^\top P \varepsilon} + m \max_j \left\{ \frac{1}{L^{j-1}} \mathfrak{b}_j \right\}
$$

so that with Young's inequality, for any $\delta > 0$,

$$
\begin{aligned}
\frac{2}{L}\varepsilon^\top P \mathscr{L}^{-1}\left(\Phi(u,\hat{\xi}) - \Phi(u,\xi)\right) &\leq \frac{2}{L}\sqrt{\varepsilon^\top P \varepsilon}\sqrt{\alpha_2}\left|\mathscr{L}^{-1}\left(\Phi(u,\hat{\xi}) - \Phi(u,\xi)\right)\right| \\
&\leq 2m\, d_\xi \sqrt{\frac{\alpha_2}{\alpha_1}}\, \frac{\tilde{\mathfrak{a}}}{L}\, V(\varepsilon) + 2m\sqrt{\alpha_2}\sqrt{\varepsilon^\top P \varepsilon}\, \max_j \left\{ \frac{1}{L^j}\mathfrak{b}_j \right\} \\
&\leq 2m\, d_\xi \sqrt{\frac{\alpha_2}{\alpha_1}}\, \frac{\tilde{\mathfrak{a}}}{L}\, V(\varepsilon) + \frac{m^2\alpha_2}{\delta} V(\varepsilon) + \delta\left(\max_j \left\{ \frac{1}{L^j}\mathfrak{b}_j \right\}\right)^2
\end{aligned}
$$

Taking $\delta = \frac{m^2\alpha_2}{\lambda}$, it follows from (B.3) that

$$
\frac{1}{L}\, \dot{\overline{V(\varepsilon)}} \leq -2\lambda\, V(\varepsilon) + \delta\left(\max_j \left\{ \frac{1}{L^j}\mathfrak{b}_j \right\}\right)^2
$$

for any L verifying

$$
L \geq \max\left\{1, \tilde{\mathfrak{a}}\, L^*\right\} \quad , \quad L^* = \frac{2m\, d_\xi}{\lambda}\sqrt{\frac{\alpha_2}{\alpha_1}} \,. \tag{B.9}
$$

This gives

$$
V(\varepsilon(t)) \leq V(\varepsilon(t_0))\, e^{-2\lambda L(t-t_0)} + \frac{\delta}{2\lambda}\left(\max_j \left\{ \frac{1}{L^j}\mathfrak{b}_j \right\}\right)^2 ,
$$

and therefore, there exists $\beta > 0$ and $\gamma > 0$ such that

$$|\varepsilon(t)| \leq \max\left\{\beta\,|\varepsilon(t_0)|\,e^{-\lambda L(t-t_0)}\ ,\ \gamma\ \max_j\left\{\frac{1}{L^j}\mathfrak{b}_j\right\}\right\}\ .$$

Because $e_i = L^{i-1}\varepsilon_i$ and $|\varepsilon| \leq |e|$, we finally get

$$|e_i(t)| \leq \max\left\{L^{i-1}\beta\,|e(t_0)|\,e^{-\lambda L(t-t_0)}\ ,\ \gamma\ \max_j\left\{L^{i-j-1}\,\mathfrak{b}_j\right\}\right\}\ .$$

From (B.8) and (B.9), the result will follow if there exist L and σ_{ij} such that

$$L > \max_{i\geq j}\left\{\mathfrak{a}_{ij}L^*, 1\right\}\ ,\quad \max_{i,j}\sum_{\ell=1}^{j}\gamma\ \mathfrak{b}_{j\ell}L^{i-j-1} \leq \varepsilon\ . \tag{B.10}$$

At this point, we have to work with the expressions of $\mathfrak{a}_{ij}$ and $\mathfrak{b}_{j\ell}$ given in (B.6). From (4.5), α_{ij} can be zero only if $i = d_\xi$. And, when $\alpha_{d_\xi\ell} = 0$, we get

$$\gamma\mathfrak{b}_{d_\xi\ell}L^{i-d_\xi-1} \ =\ \gamma\mathfrak{a}L^{i-d_\xi-1}\ \leq\ \frac{\gamma\mathfrak{a}}{L}$$

Say that we pick $\sigma_{d_\xi\ell} = 1$ in this case. For all the other cases, we choose

$$\sigma_{j\ell} = \left(\frac{2j\gamma}{\varepsilon}(1-\alpha_{j\ell})L^{(d_\xi-j-1)}\right)^{1-\alpha_{j\ell}}\ ,$$

to obtain from (B.6)

$$\gamma\mathfrak{b}_{j\ell}L^{i-j-1}\ \leq\ \varepsilon\frac{1}{j}\frac{1}{2L^{d_\xi-i}}\ .$$

So, with this selection of the $\sigma_{j\ell}$, the right inequality in (B.10) is satisfied for L sufficiently large. Then, according to (B.6), the a_{ij} are independent of L or proportional to $L^{(d_\xi-i-1)\frac{1-\alpha_{ij}}{\alpha_{ij}}}$. But with (4.5), we have

$$0 < (d_\xi - i - 1)\frac{1-\alpha_{ij}}{\alpha_{ij}} < 1\ .$$

This implies that $\frac{\mathfrak{a}_{ij}}{L}$ tends to 0 as L tends to $+\infty$. We conclude that (B.10) holds if we pick L sufficiently large. □

B.2 Homogeneous High-Gain Observer

We prove here the asymptotic convergence of the homogeneous observer (4.8) for the triangular form (4.2) verifying the homogeneous Hölder-type conditions (4.6), namely Theorem 4.3 as in [2]. In order not to over-complicate the notations, we treat the case where $d_y = 1$; namely, each block ξ_i is one-dimensional. The reader will easily convince himself that the same reasoning holds for higher output dimensions, by treating together the jth line of each block ξ_i.

As a preliminary step, we first need to build a robust homogeneous Lyapunov function for the chain of integrators without the nonlinearities.

B.2.1 Smooth Homogeneous Lyapunov Function for a Chain of Integrators

Recall the signed power function $\lfloor\cdot\rceil^b$ and the set-valued sign function S defined in (4.9).

Consider $V : \mathbb{R}^m \to \mathbb{R}_+$ be the function defined as

$$V(\bar{e}) = \sum_{i=1}^{m-1} \int_{\lfloor \bar{e}_{i+1} \rceil^{\frac{r_i}{r_{i+1}}}}^{\ell_i \bar{e}_i} \left[\lfloor \tau \rceil^{\frac{d_V - r_i}{r_i}} - \lfloor \bar{e}_{i+1} \rceil^{\frac{d_V - r_i}{r_{i+1}}} \right] d\tau + \frac{|\bar{e}_m|^{d_V}}{d_V}, \tag{B.11}$$

where d_V and ℓ_i are positive real numbers such that $d_V > 2m - 1$.

Lemma B.1 *For all d_0 in $[-1, 0]$, the function V defined in (B.11) is positive definite and there exist positive real numbers $\ell_1, \dots \ell_m$, $\tilde{\lambda}$ such that for all $\bar{e}$ in $\mathbb{R}^m$, the following holds*[5]*:*

$$\max \left\{ \frac{\partial V}{\partial \bar{e}}(\bar{e}) \left(A_m \bar{e} + \mathfrak{K}(\bar{e}_1) \right) \right\} \leq -\tilde{\lambda} V(\bar{e})^{\frac{d_V + d_0}{d_V}}, \tag{B.12}$$

with A_m the shifting matrix of order m, and $\mathfrak{K}$ defined as

$$(\mathfrak{K}(e_1))_i = -k_i \lfloor e_1 \rceil^{\frac{r_{i+1}}{r_1}} \tag{B.13}$$

where r_i is defined in (4.7) and

$$k_i = \ell_i^{\frac{ri+1}{r_i}} \ell_{i-1}^{\frac{r_{i+1}}{r_{i-1}}} \dots \ell_2^{\frac{r_{i+1}}{r_2}} \ell_1^{\frac{r_{i+1}}{r_1}}. \tag{B.14}$$

[5]Here, the max is with respect to s in $\lfloor \bar{e}_1 \rceil^0 = \mathrm{S}(\bar{e}_1)$ appearing in the mth component of $\mathfrak{K}(\bar{e}_1)$ when $d_0 = -1$.

Proof (*Case* $\mathbf{d_0} = -\mathbf{1}$ *(see [1] otherwise)*) We denote $E_i = (\bar{e}_i, \ldots, \bar{e}_m)$. Let d_V be an integer such that $d_V > 2m - 1$ and the functions $\mathfrak{K}_i$ recursively defined by:

$$\mathfrak{K}_m(\bar{e}_m) = -\lfloor \bar{e}_m \rceil^0 = -\mathrm{S}(\bar{e}_m) \quad , \quad \mathfrak{K}_i(\bar{e}_i) = \begin{pmatrix} -\lfloor \ell_i \bar{e}_i \rceil^{\frac{r_{i+1}}{r_i}} \\ \mathfrak{K}_{i+1}\left(\lfloor \ell_i \bar{e}_i \rceil^{\frac{r_{i+1}}{r_i}}\right) \end{pmatrix} ,$$

so that $\mathfrak{K}$ defined in (B.13) corresponds in fact to $\mathfrak{K}_1$. Note that the jth component of $\mathfrak{K}_i$ is homogeneous of degree $r_{j+1} = m - j$ and, for any $\bar{e}_i$ in $\mathbb{R}$, the set $\mathfrak{K}_i(\bar{e}_i)$ can be expressed as

$$\mathfrak{K}_i(\bar{e}_i) = \{\tilde{\mathfrak{K}}_i(\bar{e}_i, s) , \quad s \in \mathrm{S}(\bar{e}_i)\} ,$$

where $\tilde{\mathfrak{K}}_i : \mathbb{R} \times [-1, 1] \to \mathbb{R}$ is a continuous (single-valued) function.

Let $V_m(\bar{e}_m) = \frac{|\bar{e}_m|^{d_V}}{d_V}$, and for all i in $\{1, \ldots, m-1\}$, let also $\bar{V}_i : \mathbb{R}^2 \to \mathbb{R}$ and $V_i : \mathbb{R}^{m-i+1} \to \mathbb{R}$ be the functions defined by

$$\bar{V}_i(\nu, \bar{e}_{i+1}) = \int_{\lfloor \bar{e}_{i+1} \rceil^{\frac{r_i}{r_{i+1}}}}^{\nu} \lfloor x \rceil^{\frac{d_V - r_i}{r_i}} - \lfloor \bar{e}_{i+1} \rceil^{\frac{d_V - r_i}{r_{i+1}}} \, dx ,$$

$$V_i(E_i) = \sum_{j=m-1}^{i} \bar{V}_j(\ell_j \bar{e}_j, \bar{e}_{j+1}) + V_m(\bar{e}_m) .$$

With these definitions, the Lyapunov function V defined in (B.11) is simply $V(e) = V_1(e)$ and the homogeneous vector field $\mathfrak{K}$ defined in (B.13) $\mathfrak{K}(\bar{e}_1) = \mathfrak{K}_1(\bar{e}_1)$ with k_i verifying (B.14).

The proof of Lemma B.1 is made iteratively from $i = m$ toward 1. At each step, we show that V_i is positive definite and we look for a positive real number ℓ_i, such that for all E_i in $\mathbb{R}^{m-i+1}$

$$\max_{s \in \mathrm{S}(\bar{e}_i)} \left\{ \frac{\partial V_i}{\partial E_i}(E_i)(A_{m-i+1} E_i + \tilde{\mathfrak{K}}_i(\bar{e}_i, s)) \right\} \leq -c_i V_i(E_i)^{\frac{d_V - 1}{d_V}} , \tag{B.15}$$

where c_i is a positive real number. The lemma will be proved once we have shown that the former inequality holds for $i = 1$.

<u>Step $i = m$:</u> At this step, $E_m = \bar{e}_m$. Note that we have

$$\max_{s \in \mathrm{S}(\bar{e}_m)} \left\{ \frac{\partial V_m}{\partial E_m}(E_m) \tilde{\mathfrak{K}}_m(\bar{e}_m, s) \right\} = -|E_m|^{d_V - 1} = -c_m V_m(E_m)^{\frac{d_V - 1}{d_V}} ,$$

with $c_m = d_V^{\frac{d_V - 1}{d_V}}$. Hence, Eq. (B.15) holds for $i = m$.

<u>Step $i = j$:</u> Assume V_{j+1} is positive definite and assume there exists $(\ell_{j+1}, \ldots, \ell_m)$ such that (B.15) holds for $j = i - 1$. Note that the function $x \mapsto \lfloor x \rceil^{\frac{d_V - r_j}{r_j}} - \lfloor \bar{e}_{i+1} \rceil^{\frac{d_V - r_j}{r_{j+1}}}$ is strictly increasing, is zero if and only if $x = \lfloor \bar{e}_{j+1} \rceil^{\frac{r_j}{r_{j+1}}}$,

and therefore has the same sign as $x - \lfloor \bar{e}_{j+1} \rceil^{\frac{r_j}{r_{j+1}}}$. Thus, for any $\bar{e}_{j+1}$ fixed in $\mathbb{R}$, the function $\nu \mapsto \overline{V}_j(\nu, \bar{e}_{j+1})$ is nonnegative and is zero only for $\nu = \lfloor \bar{e}_{j+1} \rceil^{\frac{r_j}{r_{j+1}}}$. Thus, $\bar{V}_j$ is positive and we have

$$V_j(E_j) = 0 \iff \begin{cases} V_{j+1}(E_{j+1}) = 0 \\ \overline{V}_j(\ell_j \bar{e}_j, \bar{e}_{j+1}) = 0 \end{cases} \iff \begin{cases} E_{j+1} = 0 \\ \ell_j \bar{e}_j = \lfloor \bar{e}_{j+1} \rceil^{\frac{r_j}{r_{j+1}}} = 0 \end{cases}$$

so that V_j is positive definite.

On another hand, let $\tilde{V}_j(\nu, E_{j+1}) = V_{j+1}(E_{j+1}) + \bar{V}_j(\nu, \bar{e}_{j+1})$ and let T_1 be the function defined

$$T_1(\nu, E_{j+1}) = \max_{s \in \mathtt{S}(\nu)} \left\{ \tilde{T}_1(\nu, E_{j+1}, s) \right\}$$

with $\tilde{T}_1$ continuous and defined by

$$\tilde{T}_1(\nu, E_{j+1}, s) = \frac{\partial \tilde{V}_j}{\partial E_{j+1}}(E_{j+1})(A_{m-i-1}E_{i+1} + \tilde{\mathfrak{K}}_{j+1}(\lfloor \nu \rceil^{\frac{r_{j+1}}{r_j}}, s)) + \frac{c_{j+1}}{2}\tilde{V}_j$$
$$(\nu, E_{j+1})^{\frac{d_V - 1}{d_V}}.$$

Let also T_2 be the continuous real-valued function defined by

$$T_2(\nu, E_{j+1}) = -\frac{\partial \tilde{V}_j}{\partial \nu}(\nu, E_{i+1})(\bar{e}_{j+1} - \lfloor \nu \rceil^{\frac{r_{j+1}}{r_j}}).$$

Note that T_1 and T_2 are homogeneous with weight r_j for ν and r_i for $\bar{e}_i$ and degree $d_V - 1$. Besides, they verify the following two properties:

- for all E_{j+1} in $\mathbb{R}^{m-j}$, ν in $\mathbb{R}$

$$T_2(\nu, E_{j+1}) \geq 0$$

  (since $(\lfloor \nu \rceil^{\frac{r_{j+1}}{r_j}} - \bar{e}_{j+1})$ and $(\lfloor \nu \rceil^{\frac{d_V - r_j}{r_j}} - \lfloor \bar{e}_{j+1} \rceil^{\frac{d_V - r_j}{r_{j+1}}})$ have the same sign)
- for all (ν, E_{j+1}) in $\mathbb{R}^{m-j+1} \setminus \{0\}$, and s in $\mathtt{S}(\nu)$, we have the implication

$$T_2(\nu, E_{j+1}) = 0 \implies \tilde{T}_1(\nu, E_{j+1}, s) < 0$$

  since T_2 is zero only when $\lfloor \nu \rceil^{\frac{r_{j+1}}{r_j}} = \bar{e}_{j+1}$ and

$$\tilde{T}_1(\lfloor \bar{e}_{j+1} \rceil^{\frac{r_{j+1}}{r_j}}, E_{j+1}, s) = \frac{\partial V_{j+1}}{\partial E_{j+1}}(E_{j+1})(A_{n-i}E_{j+1} + \tilde{\mathfrak{K}}_{j+1}(\bar{e}_{j+1}, s))$$
$$+ \frac{c_{j+1}}{2} V_{j+1}(E_{j+1})^{\frac{d_V - 1}{d_V}} \leq -\frac{c_{j+1}}{2} V_{j+1}(E_{j+1})^{\frac{d_V - 1}{d_V}},$$

  where we have employed (B.15) for $i = j - 1$.

Using Lemma A.3 in Appendix A.1, there exists ℓ_j such that

$$T_1(\nu, E_{j+1}) - \ell_j T_2(\nu, E_{j+1}) \leq 0 \, , \; \forall \, (\nu, E_{j+1}) \, .$$

Finally, note that

$$\max_{s \in S(\bar{e}_i)} \left\{ \frac{\partial V_j}{\partial E_j}(E_j)(A_{m-j+1}E_j + \tilde{\mathfrak{K}}_j(\bar{e}_j, s)) \right\} = T_1(\ell_j \bar{e}_j) - \ell_j T_2(\ell_j \bar{e}_j, E_{j+1}) - \frac{c_{j+1}}{2} V_j(E_j)^{\frac{d_V - 1}{d_V}}$$

Hence, (B.15) holds for $i = j$. □

In order to use this Lyapunov function in the presence of nonlinearities, we need to establish its robustness.

Lemma B.2 *For all d_0 in $[-1, 0]$, the function V defined in (B.11) is positive definite and there exist positive real numbers $k_1, \ldots, k_m$, $\ell_1, \ldots, \ell_m$, λ and c_δ such that for all $\bar{e}$ in $\mathbb{R}^m$ and $\bar{\delta}$ in $\mathbb{R}^m$ the following implication*[6] *holds:*

$$|\bar{\delta}_i| \leq c_\delta V(\bar{e})^{\frac{r_{i+1}}{d_V}} \quad \forall i \qquad \Longrightarrow \qquad \max\left\{ \frac{\partial V}{\partial \bar{e}}(\bar{e})(A_m \bar{e} + \bar{\delta} + \mathfrak{K}(\bar{e}_1)) \right\} \leq -\lambda V(\bar{e})^{\frac{d_V + d_0}{d_V}}$$

Proof Let $\tilde{\mathfrak{K}}(\bar{e}_1, s)$ be the function defined as

$$\left(\tilde{\mathfrak{K}}(\bar{e}_1, s)\right)_i = (\mathfrak{K}(\bar{e}_1))_i \, , \; i \in [1, m-1] \, ,$$

and,

$$\left(\tilde{\mathfrak{K}}(\bar{e}_1, s)\right)_m = \begin{cases} k_m s \, , & \text{when } d_0 = -1 \\ (\mathfrak{K}(\bar{e}_1))_m \, , & \text{when } d_0 > -1 \end{cases} .$$

$\tilde{\mathfrak{K}}$ is a continuous (single) real-valued function which satisfies for all $\bar{e}_1$ in $\mathbb{R}$

$$\mathfrak{K}(\bar{e}_1) = \{\tilde{\mathfrak{K}}(\bar{e}_1, s) \, , \quad s \in S(\bar{e}_1)\} \, .$$

Consider also the functions

$$\tilde{\eta}(\bar{e}, \bar{\delta}, \bar{v}, s) = \frac{\partial V}{\partial \bar{e}}(\bar{e})(A_m \bar{e} + \bar{\delta} + \tilde{\mathfrak{K}}(\bar{e}_1, s)) + \frac{\tilde{\lambda}}{2} V(\bar{e})^{\frac{d_V + d_0}{d_V}} \, ,$$

and

$$\gamma(\bar{\delta}) = \sum_{i=1}^{m} |\bar{\delta}_i|^{\frac{d_V + d_0}{r_{i+1}}} \, .$$

With (B.12), we invoke Lemma A.3 to get the existence of a positive real number c_1 satisfying for all s in $S(\bar{e}_1)$:

[6] Here, the max is with respect to s in $\lfloor \bar{e}_1 \rceil^0 = S(\bar{e}_1)$ appearing in the mth component of $\mathfrak{K}(\bar{e}_1)$ when $d_0 = -1$.

$$\frac{\partial V}{\partial \bar{e}}(\bar{e})(A_m\bar{e} + \bar{\delta} + \tilde{\mathfrak{K}}(\bar{e}_1, s)) \leq -\frac{\tilde{\lambda}}{2}V(\bar{e})^{\frac{d_V+d_0}{d_V}} + c_1\sum_{i=1}^{m}\bar{\delta}_i^{\frac{d_V+d_0}{r_{i+1}}}.$$

This can be rewritten,

$$\frac{\partial V}{\partial \bar{e}}(\bar{e})(A_m\bar{e} + \bar{\delta} + \tilde{\mathfrak{K}}(\bar{e}_1, s)) \leq -\frac{\tilde{\lambda}}{2(m+1)}V(\bar{e})^{\frac{d_V+d_0}{d_V}}$$
$$+ \sum_{i=1}^{m}\left(c_1|\bar{\delta}_i|^{\frac{d_V+d_0}{r_{i+1}}} - \frac{\tilde{\lambda}}{2(m+1)}V(\bar{e})^{\frac{d_V+d_0}{d_V}}\right).$$

Consequently, the result of Lemma B.2 holds with $\lambda = \frac{\tilde{\lambda}}{2(m+1)}$, $c_\delta = \left(\frac{\lambda}{c_1}\right)^{\frac{r_1}{d_V+d_0}}$. □

B.2.2 Proof of Theorem 4.3

Proof First, the set-valued function $e_1 \mapsto \lfloor e_1 \rceil^0 = \mathtt{S}(e_1)$ defined in (4.9) is upper semi-continuous and has convex and compact values. Thus, according to [3], there exist absolutely continuous solutions to (4.8).

Let $\mathscr{L} = \mathsf{diag}(1, L, \ldots, L^{m-1})$. The error $e = \hat{\xi} - \xi$ produced by the observer (4.8) satisfies

$$\dot{e} \in LA_m e + \delta + L\mathscr{L}\mathfrak{K}(e_1) \tag{B.16}$$

where A_m is the shifting matrix of order m,

$$\delta = \Phi(u, \hat{\xi}) - \Phi(u, \xi)\,,$$

and $\mathfrak{K}$ is the homogeneous correction term defined in (B.13). In the scaled error coordinates $\varepsilon = \mathscr{L}^{-1}e$, those error dynamics read

$$\frac{1}{L}\dot{\varepsilon} \in A_m\varepsilon + \mathscr{D}_L + \mathfrak{K}(\varepsilon_1) \tag{B.17}$$

with $\mathscr{D}_L = \mathscr{L}^{-1}\delta$. We have seen in Lemma B.2 that V is an ISS Lyapunov function for the auxiliary system

$$\dot{\bar{e}} \in A_m\bar{e} + \mathfrak{K}(\bar{e}_1) \tag{B.18}$$

with state $\bar{e}$ in $\mathbb{R}^m$. See [5, Proof of Lemma 2.14]. We know need to extend it to (B.17) by a robustness analysis, and finally, deduce the result on (B.16).

Since Φ satisfies $\mathscr{H}(\alpha, \mathfrak{a})$, with (4.6) and $\frac{r_{i+1}}{r_j} \leq 1$, we obtain, for all $L \geq 1$

$$
\begin{aligned}
|\mathscr{D}_{L,i}| &\le \frac{\mathfrak{a}}{L}\sum_{j=1}^{i} L^{(j-1)\frac{r_{i+1}}{r_j}-i+1}|\varepsilon_j|^{\frac{r_{i+1}}{r_j}} \\
&\le \frac{\mathfrak{a}}{L}\sum_{j=1}^{i}|\varepsilon_j|^{\frac{r_{i+1}}{r_j}} \\
&\le \frac{c}{L}V(\varepsilon)^{\frac{r_{i+1}}{d_V}} ,
\end{aligned}
$$

where c is a positive real number obtained from Lemma A.2 in Appendix A.1. With Lemma B.2, where $\bar{\delta}_i$ plays the role of $\mathscr{D}_{L,i}$ and $\bar{e}$ the role of ε, we obtain that, by picking L^* sufficiently large such that $\frac{c}{L^*} \le \frac{c_\delta}{2}$, we have, for all $L > L^*$,

$$
\frac{1}{L}\max\left\{\frac{\partial V}{\partial e}(\varepsilon)\dot{\varepsilon}\right\} \le -\lambda V(\varepsilon)^{\frac{d_V+d_0}{d_V}} . \tag{B.19}
$$

Now, when evaluated along a solution, ε gives rise to an absolutely continuous function $t \mapsto \varepsilon(t)$. Similarly, the function defined by $t \mapsto \nu(t) = V(\varepsilon(t))$ is absolutely continuous. It follows that its time derivative is defined for almost all t and, according to [4, p. 174], (B.19) implies, for almost all t,

$$
\frac{1}{L}\dot{\nu}(t) \le -\lambda\, \nu(t)^{\frac{d_V+d_0}{d_V}} . \tag{B.20}
$$

Since with Lemma A.2, there exist a positive real number c_1 such that

$$
\left|\frac{e_i}{L^{i-1}}\right| \le c_1 V(\varepsilon)^{\frac{r_i}{d_V}} .
$$

the result follows. □

B.3 High-Gain Kalman Observer

We prove here that combining Kalman and high-gain designs gives an asymptotically stable observer (4.18) for the general triangular form (4.14).

Proof Let us start by studying the solutions to (4.16). Take any ξ_0 in $\mathbb{R}^{d_\xi}$ and denote $\nu = (u, y_{\xi_0,u})$. Observe that $LA(\nu) = \mathscr{L}^{-1}A(\nu)\mathscr{L}$, so that if Ψ_ν denotes the transition matrix[7] associated with the system $\dot{\chi} = A(\nu)\chi$, $\mathscr{L}^{-1}\Psi_\nu\mathscr{L}$ is the transition matrix associated with the system $\dot{\chi} = LA(\nu)\chi$. It follows that the solutions to (4.16) are given by

[7] See Definition 2.1.

$$P(t) = e^{-L\lambda t} \mathscr{L} \Psi_\nu^\top(0,t) \mathscr{L}^{-1} P(0) \mathscr{L}^{-1} \Psi_\nu(0,t) \mathscr{L}$$
$$+ L \int_0^t e^{-L\lambda(t-s)} \mathscr{L} \Psi_\nu^\top(\tau,t) \mathscr{L}^{-1} C(u(\tau))^\top C(u(\tau)) \mathscr{L}^{-1} \Psi_\nu(\tau,t) \mathscr{L}\, d\tau\ .$$

Therefore, $P(t)^\top = P(t)$, and because $P(0) > 0$, for all $t \geq \frac{1}{L}$,

$$P(t) \geq L \int_{t-\frac{1}{L}}^t e^{-L\lambda(t-s)} \mathscr{L} \Psi_\nu^\top(\tau,t) \mathscr{L}^{-1} C(u(\tau))^\top C(u(\tau)) \mathscr{L}^{-1} \Psi_\nu(\tau,t) \mathscr{L}\, d\tau$$

and since $C(u)\mathscr{L}^{-1} = \frac{1}{L} C(u)$,

$$\begin{aligned} P(t) &\geq \frac{1}{L} e^{-\lambda} \mathscr{L} \int_{t-\frac{1}{L}}^t \Psi_\nu^\top(\tau,t) C(u(\tau))^\top C(u(\tau)) \Psi_\nu(\tau,t) d\tau\, \mathscr{L} \\ &\geq \frac{1}{L} e^{-\lambda} \mathscr{L}\, \Gamma_\nu^b \left(t - \frac{1}{L}, t\right) \mathscr{L} \\ &\geq \alpha e^{-\lambda} I \end{aligned}$$

On the other hand, we have

$$\mathscr{L}^{-1} \Psi_\nu(\tau,t) \mathscr{L} = I + \int_\tau^t L A(\nu(s)) \mathscr{L}^{-1} \Psi_\nu(s,t) \mathscr{L} ds$$

so that for all $\tau \leq t$,

$$|\mathscr{L}^{-1} \Psi_\nu(\tau,t) \mathscr{L}| = 1 + \int_\tau^t L A_{max} |\mathscr{L}^{-1} \Psi_\nu(s,t) \mathscr{L}| ds$$

and by Gronwall's lemma,

$$|\mathscr{L}^{-1} \Psi_\nu(\tau,t) \mathscr{L}| = e^{L A_{max}(t-\tau)}\ .$$

It follows from the expression of P that

$$\begin{aligned} P(t) &\leq e^{-L(\lambda - 2A_{max})t} |P(0)| + L \int_0^t e^{-L(\lambda - 2A_{max})(t-\tau)} |C(u(\tau))^\top C(u(\tau))| d\tau \\ &\leq |P(0)| + \frac{C_{max}^2}{\lambda - 2A_{max}} \end{aligned}$$

for $\lambda > 2A_{max}$. We conclude that for any positive definite matrix P_0, there exist positive scalars α_1 and α_2 such that for any ξ_0 in $\mathbb{R}^{d_\xi}$, for any $\lambda > 2A_{max}$ and $L \geq L_0$, the solution to (4.16) initialized at P_0 verifies (4.17).

Consider now a positive definite matrix P_0 giving (4.17) for any ξ_0 in $\mathbb{R}^{d_\xi}$, any $\lambda > 2A_{max}$ and any $L \geq L_0$. Consider initial conditions ξ_0 and $\hat{\xi}_0$ in $\mathbb{R}^{d_\xi}$ for systems

(4.14) and (4.18), respectively, and the corresponding solution $t \mapsto P(t)$ to (4.16). Define the function

$$V(\xi, \hat{\xi}, t) = (\hat{\xi} - \xi)^\top \mathscr{L}^{-1} P(t) \mathscr{L}^{-1} (\hat{\xi} - \xi) \,.$$

Thanks to (4.17), it verifies for all $(\xi, \hat{\xi})$ in $\mathbb{R}^{d_\xi} \times \mathbb{R}^{d_\xi}$, and all $t \geq \frac{1}{L}$,

$$\alpha_1 |\hat{\xi} - \xi|^2 \leq V(\xi, \hat{\xi}, t) \leq \alpha_2 |\hat{\xi} - \xi|^2 \,. \tag{B.21}$$

Also, according to (4.14) and (4.18),

$$\dot{\overbrace{\hat{\xi} - \xi}} = \Big(A(u, y) - \mathscr{L} P^{-1} C(u)^\top C(u)\Big)(\hat{\xi} - \xi) + \Big(\Phi(u, \hat{\xi}) - \Phi(u, \xi)\Big)$$

and

$$\mathscr{L}^{-1} \dot{\overbrace{\hat{\xi} - \xi}} = L\Big(A(u, y) - P^{-1} C(u)^\top C(u)\Big)\mathscr{L}^{-1}(\hat{\xi} - \xi) + \mathscr{L}^{-1}\Big(\Phi(u, \hat{\xi}) - \Phi(u, \xi)\Big)$$

so that (omitting the dependence on t to ease the notations)

$$\begin{aligned}
\dot{\overbrace{V(\xi, \hat{\xi}, t)}} &= (\hat{\xi} - \xi)^\top \mathscr{L}^{-1} \left[\dot{P} + L\left(P A(u, y) - C(u)^\top C(u)\right)\right. \\
&\qquad \left. + L\left(A(u, y) \top P - C(u)^\top C(u)\right)\right] \mathscr{L}^{-1}(\hat{\xi} - \xi) \\
&\qquad + 2(\hat{\xi} - \xi)^\top \mathscr{L}^{-1} P \mathscr{L}^{-1}\Big(\Phi(u, \hat{\xi}) - \Phi(u, \xi)\Big) \\
&= -L\lambda\, V(\xi, \hat{\xi}, t) - L(\hat{\xi} - \xi)^\top \mathscr{L}^{-1} C(u)^\top C(u) \mathscr{L}^{-1}(\hat{\xi} - \xi) \\
&\qquad + 2(\hat{\xi} - \xi)^\top \mathscr{L}^{-1} P \mathscr{L}^{-1}\Big(\Phi(u, \hat{\xi}) - \Phi(u, \xi)\Big) \\
&\leq -L\lambda\, V(\xi, \hat{\xi}, t) + 2(\hat{\xi} - \xi)^\top \mathscr{L}^{-1} P \mathscr{L}^{-1}\Big(\Phi(u, \hat{\xi}) - \Phi(u, \xi)\Big) \,.
\end{aligned}$$

Thanks to the triangularity and Lipschitzness of Φ, using the same arguments as in the proof of Theorem 4.1, we get for $L \geq 1$, and all $t \geq \frac{1}{L}$,

$$\dot{\overbrace{V(\xi, \hat{\xi}, t)}} \leq -L\left(\lambda - 2m \frac{\alpha_2}{\alpha_1} \frac{\mathfrak{a}}{L}\right) V(\xi, \hat{\xi}, t) \,.$$

It follows that for

$$L \geq \max\left\{1, \frac{4m}{\lambda} \frac{\alpha_2}{\alpha_1} \mathfrak{a}\right\} ,$$

and all $t \geq \frac{1}{L}$,

$$V(\xi, \hat{\xi}, t) \leq e^{-L \frac{\lambda}{2}(t - \frac{1}{L})}\, V\left(\xi(\frac{1}{L}), \hat{\xi}(\frac{1}{L}), \frac{1}{L}\right)$$

which gives the result with (B.21). □

References

1. Andrieu, V., Praly, L., Astolfi, A.: Homogeneous approximation, recursive observer design, and output feedback. SIAM J. Control Optim. **47**(4), 1814–1850 (2008)
2. Bernard, P., Praly, L., Andrieu, V.: Observers for a non-Lipschitz triangular form. Automatica **82**, 301–313 (2017)
3. Filippov, A.: Differential Equations with Discontinuous Right-Hand Sides. Mathematics and Its Applications. Kluwer Academic Publishers Group, Dordrecht (1988)
4. Smirnov, G.: Introduction to the Theory of Differential Inclusions. Graduate Studies in Mathematics, vol. 41. American Mathematical Society, Providence (2001)
5. Sontag, E., Wang, Y.: On characterizations of the input-to-state stability property. Syst. Control Lett. **24**, 351–359 (1995)

Appendix C
Injectivity Analysis for Nonlinear Luenberger Designs

This appendix aims at proving some results of Chap. 6. More precisely, we show how to prove the injectivity of a map T solution of a PDE

$$\frac{\partial T}{\partial x}(x) f(x) = A\,T(x) + B\,h(x)\ ,$$

for an autonomous system, or more generally

$$\frac{\partial T}{\partial x}(x,t) f(x,u(t)) + \frac{\partial T}{\partial t}(x,t) = A\,T(x,t) + B\,h(x,u(t))$$

in the general case, with $A \in \mathbb{R}^{d_\xi \times d_\xi}$ Hurwitz and $B \in \mathbb{R}^{d_\xi \times d_y}$ such that the pair (A, B) is controllable, when some observability assumptions are satisfied. This roughly says that if the output (given by h) is sufficiently "rich" in terms of x throughout time, filtering it at d_ξ distinct (maybe sufficiently large) frequencies (given by A) gives a quantity that contains enough information to reconstruct x at each time (or at least after a certain time if T is time-varying). This is typically done under two types of observability:

- *strong differential observability* such as in [1, Theorem 4] for autonomous systems and in [2, Theorem 2] for non-autonomous systems (recalled in Theorem 6.4 of this book). In that case, the map H made of the outputs h_i and a certain number m_i of their derivatives is an injective immersion (see Assumption 6.2) and this property can be transferred to the map T, by taking T of the same dimension as H and exploiting the fact that for large eigenvalues of A, T is roughly a linear combination of the components of H.
- *backward distinguishability* in $\mathscr{X}$ such as in [1, Theorem 3] for autonomous systems (Theorem 6.3 of this book) and in [3, Theorem 3] for non-autonomous systems (Theorem 6.5 of this book). In that weaker case, one can show that T is injective for almost any choice of $d_x + 1$ complex eigenvalues of A.

P. Bernard, *Observer Design for Nonlinear Systems*,
Lecture Notes in Control and Information Sciences 479,
https://doi.org/10.1007/978-3-030-11146-5

C.1 Based on Strong Differential Observability

We prove here Theorem 6.4 as in [2] with a time-varying T for possibly non-autonomous systems.

Proof Since A and B are taken of the form

$$A = \begin{pmatrix} kA_1 & & & & \\ & \ddots & & & \\ & & kA_i & & \\ & & & \ddots & \\ & & & & kA_{d_y} \end{pmatrix} \qquad B = \begin{pmatrix} B_1 & & & & \\ & \ddots & & & \\ & & B_i & & \\ & & & \ddots & \\ & & & & B_{d_y} \end{pmatrix},$$

the map T can be decomposed as

$$T(x,t) = \big(T_1(x,t), \dots, T_i(x,t), \dots, T_{d_y}(x,t)\big) \tag{C.1}$$

with

$$\frac{\partial T_i}{\partial x}(x,t) f(x,u(t)) + \frac{\partial T_i}{\partial t}(x,t) = kA_i\, T_i(x,t) + B_i\, h_i(x,u(t))\,. \tag{C.2}$$

Take u in $\mathscr{U}$, i in $\{1, \dots, d_y\}$, x in $\mathscr{S}$ and t in $[0, +\infty)$. According to PDE (C.2), T_i satisfies for all s in $[0, +\infty)$,

$$\frac{d}{ds} T_i(X(x,t;s;u), s) = kA_i\, T_i(X(x,t;s;u), s) + B_i Y_i(x,t;s;u)\,.$$

Integrating between t and s, it follows that

$$T_i(X(x,t;s;u), s) = e^{kA_i(s-t)} \underbrace{T_i(X(x,t;t;u), t)}_{T_i(x,t)} + \int_t^s e^{kA_i(s-\tau)} B_i\, Y_i(x,t;\tau;u)d\tau$$

and thus,

$$T_i(x,t) = e^{kA_i(t-s)}\, T_i(X(x,t;s;u), s) + \int_s^t e^{kA_i(t-\tau)} BY_i(x,t;\tau;u)d\tau\,.$$

Applying this inequality at $s = 0$, we get

$$T_i(x,t) = e^{kA_i t}\, T_i(X(x,t;0;u), 0) + T_i^0(x,t)$$

where T_i^0 is such that T^0 defined in (6.8) is

$$T^0(x,t) = \left(T_1^0(x,t), \ldots, T_i^0(x,t), \ldots, T_{d_y}^0(x,t)\right) .$$

But after m_i successive integration by parts in (6.8), we get,

$$T_i^0(x,t) = -A_i^{-m_i}\mathscr{C}_i\, K_i\, H_i(x, \overline{u}_m(t)) \\ + A_i^{-m_i} e^{kA_i t}\mathscr{C}_i K_i\, H_i(X(x,t;0;u), \overline{u}_m(0)) + \frac{1}{k^{m_i}} A_i^{-m_i} R_i(x,t)$$

where $K_i = \text{diag}\left(\frac{1}{k}, \ldots, \frac{1}{k^{m_i}}\right)$, $\mathscr{C}_i$ is the invertible controllability matrix

$$\mathscr{C}_i = [A_i^{m_i-1} B_i \,, \ldots, A_i B_i \,, B_i] \,,$$

$H_i(x, \overline{\nu}_m)$ is defined in (6.10), and R_i is the remainder:

$$R_i(x,t) = \int_0^t e^{kA_i(t-\tau)} B_i\, L_{\bar{f}}^{m_i} h_i\, (X(x,t;\tau;u), \overline{u}_m(\tau))\, d\tau$$

We finally deduce that

$$T_i(x,t) = A_i^{-m_i}\,\mathscr{C}_i\, K_i \left(-H_i(x, \overline{u}_m(t)) + K_i^{-1}\mathscr{C}_i^{-1}\left(e^{kA_i t}\,\Psi_i(X(x,t;0;u),0) + \frac{1}{k^{m_i}} R_i(x,t)\right)\right)$$

with $\Psi_i(x,t) = A_i^{m_i} T_i(x,t) + \mathscr{C}_i K_i\, H_i(x, \overline{u}_m(t))$.

Let us now consider x_1 and x_2 in $\mathscr{S}$, and t in $[0, +\infty)$. We are interested in the quantity $|T(x_1,t) - T(x_2,t)|$, and thus in $|T_i(x_1,t) - T_i(x_2,t)|$.

Thanks to Assumption 6.2.2, for any (x_1, x_2) in $\mathscr{S}$, and (t, τ) in $[0, +\infty)^2$, we have (see for instance [4])

$$|X(x_1,t;\tau;u) - X(x_2,t;\tau;u)| \leq e^{M_f|\tau - t|}|x_1 - x_2| \,. \tag{C.3}$$

By assumption $T_i(\cdot, 0)$ and $H_i(\cdot, \overline{u}_m(0)))$ are Lipschitz on $\mathscr{S}$, thus there exists L_{Ψ_i} such that

$$|\Psi_i(X(x_1,t;0;u),0) - \Psi_i(X(x_2,t;0;u),0)| \leq L_{\Psi_i}\, e^{M_f t}|x_1 - x_2| \,.$$

Then, A_i being Hurwitz, there exists strictly positive numbers a_i and γ_i (see [4]) such that for all τ in $[0, t]$

$$\left|e^{kA_i(t-s)}\right| \leq \gamma_i\, e^{-ka_i(t-s)} \,. \tag{C.4}$$

Using Assumption 6.2.4 and inequalities (C.3) and (C.4), we deduce that if $k > \dfrac{M_f}{a_i}$,

$$|R_i(x_1,t) - R_i(x_2,t)| \leq L_i|B_i|\gamma_i \int_0^t e^{-(ka_i - M_f)(t-\tau)} d\tau\, |x_1 - x_2| \leq \frac{L_i|B_i|\gamma_i}{ka_i - M_f}\, |x_1 - x_2| \,.$$

We finally deduce that

$$
\begin{aligned}
|T_i(x_1,t) - T_i(x_2,t)| &\geq |A_i^{-m_i}\,\mathscr{C}_i\,K_i| \left(|\Delta H_i| - |K_i^{-1}\mathscr{C}_i^{-1}| \left(\left|e^{kA_i t}\right| |\Delta\Psi_i| + \frac{1}{k^{m_i}} |\Delta R_i| \right) \right) \\
&\geq \frac{|A_i^{-m_i}\,\mathscr{C}_i|}{k^{m_i}} \Big(\, |\Delta H_i| - k^{m_i} |\mathscr{C}_i^{-1}|\,\gamma_i\,L_{\Psi_i}\,e^{-(ka_i - M_f)t}\,|x_1 - x_2| \\
&\qquad\qquad - |\mathscr{C}_i^{-1}|\,\gamma_i \frac{L_i |B_i|}{ka_i - M_f}\,|x_1 - x_2| \Big)
\end{aligned}
$$

where ΔH_i, $\Delta\Psi_i$, and ΔR_i denote the difference of the functions $H_i(\cdot, \overline{u}_m(t))$, $\Psi_i(X(\cdot, t; 0; u), 0)$, and $R_i(\cdot, t)$, respectively, evaluated at x_1 and x_2. It follows (by norm equivalence) that there exists a constant c such that

$$
\begin{aligned}
|T(x_1,t) - T(x_2,t)| \geq c\, &\frac{\min_i(|A_i^{-m_i}\,\mathscr{C}_i|)}{k^m} \Big[\, |H(x_1, \overline{u}_m(t)) - H(x_2, \overline{u}_m(t))| \\
&- \left(\left(\sum_{i=1}^{p} k^{m_i}\,\gamma_i |\mathscr{C}_i^{-1}|\,L_{\Psi_i} \right) e^{-(ka - M_f)t} + \frac{\left(\sum_{i=1}^{p} L_i\gamma_i |\mathscr{C}_i^{-1}||B_i|\right)}{ka - M_f} \right) |x_1 - x_2| \Big] \\
\geq c\, &\frac{\min_i(|A_i^{-m_i}\,\mathscr{C}_i|)}{k^m} \left(\frac{1}{L_H} - c_1\,k^m\,e^{-(ka - M_f)t} - c_2 \frac{1}{ka - M_f} \right) |x_1 - x_2|
\end{aligned}
$$

where m a, c_1, and c_2 are constants independent from k and t defined by

$$
m = \max_i m_i \quad , \quad a = \min_i a_i \quad , \quad c_1 = \sum_{i=1}^{p} \gamma_i |\mathscr{C}_i^{-1}|\,L_{\Psi_i} \quad , \quad c_2 = \sum_{i=1}^{p} L_i\gamma_i |\mathscr{C}_i^{-1}||B_i|\,.
$$

We deduce that for

$$
k \geq \frac{1}{a}\left(M_f + 4c_2 L_H\right) \quad , \quad t \geq \frac{\ln(4k^m c_1 L_H)}{ka - M_f}\,,
$$

we have

$$
|T(x_1,t) - T(x_2,t)| \geq c\frac{\min_i(|A_i^{-m_i}\,\mathscr{C}_i|)}{k^m}\,\frac{1}{2L_H}\,|x_1 - x_2|
$$

i.e.,

$$
|x_1 - x_2| \leq 2L_H \frac{k^m}{c\,\min_i(|A_i^{-m_i}\,\mathscr{C}_i|)}\,|T(x_1,t) - T(x_2,t)| \tag{C.5}
$$

and $T(\cdot, t)$ is injective on $\mathscr{S}$, uniformly in time. We conclude that the result holds with

$$
\overline{k} = \frac{1}{a}\left(M_f + 4c_2 L_H\right) \quad , \quad L_k = 2L_H \frac{k^m}{c\,\min_i(|A_i^{-m_i}\,\mathscr{C}_i|)}
$$

$$
\overline{t}_{k,u} = \max\left\{ \frac{\ln(4k^m c_1 L_H)}{ka - M_f}\,,\, 0 \right\}.
$$

Since M_f, L_H, and L_i (and thus c_2) are independent from u, $\overline{k}$, and L_k are the same for all u in $\mathscr{U}$, while $\overline{t}_{k,u}$ depends on u through L_{Ψ_i}.

Now, take any x in $\mathscr{S}$ and $t \geq \overline{t}_{k,u}$. For any v and any h such that $x + hv$ is in $\mathscr{S}$, we have

$$L_k|v| \leq \frac{|T(x+hv,t) - T(x,t)|}{|h|}$$

and by letting h go to 0, we get

$$L_k|v| \leq \left|\frac{\partial T}{\partial x}(x,t)v\right| .$$

Hence, $T(\cdot,t)$ is an immersion on $\mathscr{S}$. □

C.2 Based on Backward Distinguishability

We prove here Theorem 6.5 as in [3] with a time-varying T for possibly non-autonomous systems.

Proof Let us define for λ in $\mathbb{C}$, the function $T^0_\lambda : \mathscr{S} \times \mathbb{R}^+ \to \mathbb{C}^{d_y}$

$$T^0_\lambda(x,t) = \int_0^t e^{-\lambda(t-s)} Y(x,t;s;u)\, ds . \tag{C.6}$$

Given the structure of A and B, and with a permutations of the components,

$$T^0(x,t) = \left(T^0_{\lambda_1}(x,t), \ldots, T^0_{\lambda_{d_x+1}}(x,t)\right) .$$

We need to prove that T^0 is injective for almost all $(\lambda_1, \ldots, \lambda_{d_x+1})$ in Ω^{d_x+1} (in the sense of the Lebesgue measure). For that, we define the function

$$\Delta T(x_a,x_b,t,\lambda) = T^0_\lambda(x_a,t) - T^0_\lambda(x_b,t)$$

on $\Upsilon \times \Omega$ with

$$\Upsilon = \{(x_a,x_b,t) \in \mathscr{S}^2 \times (\overline{t}_u, +\infty) \ : \ x_a \neq x_b\} .$$

We are going to use the following lemma whose proof[8] can be found in [1]: □

[8]More precisely, the result proved in [1] is for Υ open set of $\mathbb{R}^{2d_x}$ instead of $\mathbb{R}^{2d_x+1}$. But the proof turns out to be still valid with $\mathbb{R}^{2d_x+1}$ because the only constraint is that the dimension of Υ be strictly less than $2(d_\xi + 1)$.

Lemma C.1 (Coron's lemma) *Let Ω and Υ be open sets of $\mathbb{C}$ and $\mathbb{R}^{2d_x+1}$, respectively. Let $\Delta T : \Upsilon \times \Omega \to \mathbb{C}^{d_y}$ be a function which is holomorphic in λ for all $\underline{x}$ in Υ and C^1 in $\underline{x}$ for all λ in Ω. If for any $(\underline{x}, \lambda)$ in $\Upsilon \times \Omega$ such that $\Delta T(\underline{x}, \lambda) = 0$, there exists i in $\{1, \ldots, d_y\}$ and $k > 0$ such that $\frac{\partial^k \Delta T_i}{\partial \lambda^k}(\underline{x}, \lambda) \neq 0$, then the set*

$$\mathscr{R} = \bigcup_{\underline{x} \in \Upsilon} \left\{(\lambda_1, \ldots, \lambda_{d_x+1}) \in \Omega^{d_x+1} \ : \ \Delta T(\underline{x}, \lambda_1) = \ldots = \Delta T(\underline{x}, \lambda_{d_x+1}) = 0\right\}$$

has zero Lebesgue measure in $\mathbb{C}^{d_x+1}$.

In our case, ΔT is clearly holomorphic in λ and C^1 in $\underline{x}$. Since for every $\underline{x}$ in Υ, $\lambda \mapsto \Delta T(\underline{x}, \lambda)$ is holomorphic on the connex set $\mathbb{C}$, we know that its zeros are isolated and admit a finite multiplicity, unless it is identically zero on $\mathbb{C}$. In the latter case, we have in particular for any ω in $\mathbb{R}$

$$\int_{-\infty}^{+\infty} e^{-i\omega\tau} g(\tau)\, d\tau = 0$$

with g the function

$$g(\tau) = \begin{cases} Y(x_a, t; t - \tau; u) - Y(x_b, t; t - \tau; u) & , \text{ if } \tau \in [0, t] \\ 0 & , \text{ otherwise} \end{cases}$$

which is in $\mathscr{L}^2$. Thus, the Fourier transform of g is identically zero and we deduce that necessarily

$$Y(x_a, t; t - \tau; u) - Y(x_b, t; t - \tau; u) = 0$$

for almost all τ in $[0, t]$ and thus for all τ in $[0, t]$ by continuity. Since $t \geq \overline{t}_u$, it follows from the backward distinguishability that $x_a = x_b$, but this is impossible because (x_a, x_b, t) is in Υ. We conclude that $\lambda \mapsto \Delta T(\underline{x}, \lambda)$ is not identically zero on $\mathbb{C}$ and the assumptions of the lemma are satisfied. Thus, $\mathscr{R}$ has zero measure, and for all $(\lambda_1, \ldots, \lambda_{d_x+1})$ in $\mathbb{C}^{d_x+1} \setminus \mathscr{R}$, T^0 is injective on $\mathscr{S}$, by definition of $\mathscr{R}$. □

References

1. Andrieu, V., Praly, L.: On the existence of a Kazantzis–Kravaris/Luenberger observer. SIAM J. Control Optim. **45**(2), 432–456 (2006)
2. Bernard, P.: Luenberger observers for nonlinear controlled systems. In: IEEE Conference on Decision and Control (2017)
3. Bernard, P., Andrieu, V.: Luenberger observers for non autonomous nonlinear systems. IEEE Trans. Autom. Control **64**(1), 270–281 (2019)
4. Miller, R., Michel, A.: Ordinary Differential Equations. Academic, New York (1982)

Index

P. Bernard, *Observer Design for Nonlinear Systems*,
Lecture Notes in Control and Information Sciences 479,
https://doi.org/10.1007/978-3-030-11146-5

MIX
Papier aus verantwortungsvollen Quellen
Paper from responsible sources
FSC® C105338

If you have any concerns about our products,
you can contact us on
ProductSafety@springernature.com

In case Publisher is established outside the EU,
the EU authorized representative is:
Springer Nature Customer Service Center GmbH
Europaplatz 3, 69115 Heidelberg, Germany

Printed by Libri Plureos GmbH
in Hamburg, Germany